Reinforcement Learning Aided Performance Optimization of Feedback Control Systems

Changsheng Hua

Reinforcement Learning Aided Performance Optimization of Feedback Control Systems

Springer Vieweg

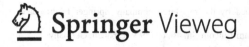

Changsheng Hua
Duisburg, Germany

Von der Fakultät für Ingenieurwissenschaften, Abteilung Elektrotechnik und Informationstechnik der Universität Duisburg-Essen zur Erlangung des akademischen Grades Doktor der Ingenieurwissenschaften (Dr.-Ing.) genehmigte Dissertation von Changsheng Hua aus Jiangsu, V.R. China.
1. Gutachter: Prof. Dr.-Ing. Steven X. Ding
2. Gutachter: Prof. Dr. Yuri A.W. Shardt
Tag der mündlichen Prüfung: 23.01.2020

ISBN 978-3-658-33033-0 ISBN 978-3-658-33034-7 (eBook)
https://doi.org/10.1007/978-3-658-33034-7

Responsible Editor: Stefanie Eggert
This Springer Vieweg imprint is published by the registered company Springer Fachmedien Wiesbaden GmbH part of Springer Nature.
The registered company address is: Abraham-Lincoln-Str. 46, 65189 Wiesbaden, Germany

To my parents, my brother, and my wife Xiaodi

Acknowledgments

First and foremost, I am deeply indebted to my advisor, Prof. Dr.-Ing. Steven X. Ding, for his guidance, encouragement and insightful discussions during my Ph.D. studies. He has continually pointed me in the right direction and provided me inspiration to do the best work I could. I would also like to express my heartfelt appreciation to Prof. Dr. Yuri A.W. Shardt, who has sparked my interest in performance optimization of control systems and mentored me a lot. He has shared with me his rich experience in academic research and scientific writing. I feel very lucky to have him as one of my major collaborators.

I would particularly like to give thanks to Dr.-Ing. Linlin Li and Dr.-Ing. Hao Luo for many insightful discussions and constructive comments during my studies. From both of them, I have learned a lot on robust and optimal control. I would also like to thank Dr.-Ing. Birgit Köppen-Seliger, Dr.-Ing. Chris Louen, Dr.-Ing. Minjia Krüger and Dr.-Ing. Tim Könings for giving me support and valuable advice in supervision of exercises and research projects.

I owe a great debt of gratitude to Dr.-Ing. Zhiwen Chen, Dr.-Ing. Kai Zhang, Dr.-Ing. Yunsong Xu, Dr.-Ing. Lu Qian, who offered enormous help and support during the early days of my life in Duisburg. I am particularly thankful to M.Sc. Micha Obergfell, M.Sc. Yuhong Na, M.Sc. Frederick Hesselmann, M.Sc. Hogir Rafiq, M.Sc. Ting Xue, M.Sc. Deyu Zhang, M.Sc. Reimann Christopher, M.Sc. Caroline Zhu, M.Sc. Yannian Liu, M.Sc. Tieqiang Wang, M.Sc. Jiarui Zhang for making my life in AKS much enjoyable, and for giving me valuable suggestions, generous support and encouragement. I would like to extend my thanks to numerous visiting scholars for all the advice, support, help and the great times. I am also very grateful for the administrative and technical assistance given by Mrs. Sabine Bay, Dipl.-Ing. Klaus Göbel and Mr. Ulrich Janzen.

Lastly, I would like to dedicate this work to my family, to my parents for their unconditional love and care, to my brother for igniting my passion for engineering and all the years of care and support, and especially to my dear wife Xiaodi for being with me all these years with patience and faithful support.

Contents

Abbreviations and Notation

Abbreviations

Abbreviation	Description
BLDC	brushless direct current
DP	dynamic programming
ECU	electronic control unit
I/O	input/output
IOR	input and output recovery
KL	Kullback-Leibler
LCF	left coprime factorization
LQG	linear quadratic Gaussian
LQR	linear quadratic regulator
LS	least squares
LTI	linear time-invariant
LTR	loop transfer recovery
MIMO	multiple-input multiple-output
NAC	natural actor-critic
PI	proportional-integral
PID	proportional-integral-derivative
RCF	right coprime factorization
RL	reinforcement learning
SARSA	state-action-reward-state-action
SGD	stochastic gradient descent
SISO	single-input single-output
TD	temporal difference

2-DOF	two-degree-of-freedom
YK	Youla-Kučera

Notation

Notation	Description
$\forall$	for all
$\in$	belong to
$\sim$	follow
$\approx$	approximately equal
$\neq$	not equal
$:=$	defined as
$\Rightarrow$	imply
$\gg$	much greater than
$\otimes$	Kronecker product
$\lvert \cdot \rvert$	determinant of a matrix or absolute value
$\mathbb{Z}^+$	set of non-negative integers
$\mathbb{R}^n$	space of n-dimensional column vectors
$\mathbb{R}^{n \times m}$	space of n by m matrices
x	a scalar
$\ln x$	natural logarithm of x
$\boldsymbol{x}$	a vector
$\boldsymbol{X}$	a matrix
$\boldsymbol{X}^T$	transpose of $\boldsymbol{X}$
$\boldsymbol{X}^{-1}$	inverse of $\boldsymbol{X}$
$\mathrm{tr}(\boldsymbol{X})$	trace of $\boldsymbol{X}$
$\boldsymbol{X} > \boldsymbol{0}$	$\boldsymbol{X}$ is a positive definite matrix
$\mathrm{vec}(\boldsymbol{X})$	vectorization of $\boldsymbol{X}$, $\mathrm{vec}(\boldsymbol{X}) = \begin{bmatrix} \boldsymbol{x}_1 \\ \vdots \\ \boldsymbol{x}_m \end{bmatrix} \in \mathbb{R}^{nm}$, for $\boldsymbol{X} = [\boldsymbol{x}_1 \cdots \boldsymbol{x}_m] \in \mathbb{R}^{n \times m}$, $\boldsymbol{x}_i \in \mathbb{R}^n, i = 1, \cdots, m$
$\boldsymbol{I}_m$	m by m identity matrix
$\boldsymbol{0}_{m \times n}$	m by n zero matrix
$\mathcal{RH}_\infty$	set of all proper and real rational stable transfer matrices
$\mathcal{R}_p$	set of all proper and real rational transfer matrices
$\mathcal{R}_{sp}$	set of all strictly proper and real rational transfer matrices
$\lVert \boldsymbol{P} \rVert_\infty$	$\mathcal{H}_\infty$ norm of a transfer matrix $\boldsymbol{P}$

$\left[\begin{array}{c\|c} A & B \\ \hline C & D \end{array}\right]$	shorthand for state-space realization $\boldsymbol{C}(z\boldsymbol{I}-\boldsymbol{A})^{-1}\boldsymbol{B}+\boldsymbol{D}$
$\arg\min_{\boldsymbol{u}}(f(\boldsymbol{u}))$	a value of $\boldsymbol{u}$ at which $f(\boldsymbol{u})$ takes its minimum value
$N(\boldsymbol{a}, \Sigma)$	Gaussian distribution with a mean vector $\boldsymbol{a}$ and a covariance matrix Σ
$\mathbb{E}(\cdot)$	Mean value/vector
μ/π	deterministic/stochastic policy
μ^*/π^*	(sub)optimal deterministic/stochastic policy
γ	discount factor
$c(\boldsymbol{x}, \boldsymbol{u})$	one-step cost
μ^i	i^{th} iteration of policy μ
$\boldsymbol{R}_j^{\mu^i}$	j^{th} iteration of a parameter matrix $\boldsymbol{R}$ under the ith iteration of the policy μ
$\pi_{\boldsymbol{\theta}}$	a stochastic policy corresponding to a parameter vector $\boldsymbol{\theta}$
$\delta(k)$	temporal difference error at the sampling instant k
$V^{\pi}(\boldsymbol{x})$	value function of policy π
$Q^{\pi}(\boldsymbol{x}, \boldsymbol{u})$	Q-function of policy π
$A^{\pi}(\boldsymbol{x}, \boldsymbol{u})$	advantage function of policy π
$\nabla_{\boldsymbol{\theta}} f(\boldsymbol{\theta})$	derivative of $f(\boldsymbol{\theta})$ with respect to a parameter vector $\boldsymbol{\theta}$

List of Figures

Introduction

1

1.1 Motivation

This thesis deals with reinforcement learning aided performance optimization of feedback control systems. The motivation of this work will be elaborated consecutively.

Why Performance Optimization of Feedback Control Systems?
Feedback control systems have been receiving more and more attention in modern industry, as they appeal to ever increasing industrial demands on higher production quality and economical benefits thanks to the following two advantages:

- They add stability to systems, and ensure reliable control loops and high control performance.
- They provide continuous control of plants and maximize their productivity, and meanwhile reduce human involvement and avoid human-related errors and unnecessary labor costs.

Therefore, feedback controllers, such as proportional-integral-derivative (PID) controllers and model predictive controllers, are widely applied to a variety of systems like aircrafts, vehicles, robots or chemical systems.

However, in the industrial practice, on the one hand, the feedback controllers are often conservatively tuned in the commissioning stage only to ensure system stability [36]. This can lead to sluggish system behavior. On the other hand, system performance can degrade, over time, on account of diverse changes, such as changes in operational conditions and replacement of components in maintenance actions. The performance degradation, if not dealt with properly, can propagate through the

C. Hua, *Reinforcement Learning Aided Performance Optimization of Feedback Control Systems*, https://doi.org/10.1007/978-3-658-33034-7_1

entire system [9], resulting in huge losses. Despite this, the main reasons for the lack of controller tuning by engineers for performance optimization during system operation in industry are:

- There are limited time and resources. Most engineers are fully busy with keeping systems in operation. They have very little time for improving controllers. Moreover, a remarkable number of controllers are maintained by few control engineers, who survive from decades of downsizing and outsourcing [36].
- There is a shortage of skilled personnel. Plant operators and engineers do not have necessary skills to ascertain causes of poor system performance. Mostly, the poor system performance is accepted as "normal" [36].
- The modification of existing controllers is risky and expensive in terms of man-hours, operation stop, and commissioning of the new system. This can lower economical benefits [6].

Strongly motivated by these observations and driven by industrial needs for high-quality productions and maximum profits, this thesis is dedicated to the study of data-driven performance optimization of feedback control systems with minimum human intervention and without modifications of existing controllers.

Why Reinforcement Learning?

Reinforcement learning (RL) is a branch of machine learning concerned with how agents take actions by interacting with their environment to maximize some notion of cumulative reward. It is a powerful tool to solve optimization problems in a data-driven manner [20, 48, 59, 65]. There have been a lot of remarkable results over the last few decades about applications of RL to performance optimization such as playing Atari games [49] and Go [62, 64] with super-human performance, and training robotics to learn primitive skills [34, 38, 39]. Meanwhile, there have been also a lot of studies on applications of RL to controller design in the control community. This includes learning of state feedback controllers [10, 41] and $\mathcal{H}_\infty$ controllers [1] for linear systems, learning of optimal controllers for nonlinear systems [28, 31]. All of them suggest the potential of RL in the application of performance optimization of feedback control systems.

Why Reinforcement Learning Aided Performance Optimization of Feedback Control Systems?

To deal with performance degradation, fault-tolerant control methods have been widely developed [9, 14, 83]. Generally, they fall into two categories 1) model-based, 2) data-driven. One kind of model-based methods is to pre-design a set of controllers for a group of faulty scenarios [51], so that once any of them is detected, the corresponding controller is switched into operation to ensure stability and maintain acceptable performance. The other kind of model-based methods is to design adaptive controllers, such as model following [82] or sliding mode controllers [44], whose parameters change in accord with plant changes, so that performance degradation is not "visible" from the plant input/output (I/O) responses. However, due to their innate model-based nature, these performance optimization methods cannot be easily transplanted among different systems.

Unlike model-based methods, data-driven ones are flexible. They require no prior knowledge about the plant model for dealing with performance degradation, since methods, such as subspace identification methods [56], can be used to extract a model directly from plant I/O data. Based on the identified model, model-based design techniques [13] can be further applied to reconfigure the existing controller to enhance system performance. However, the challenge of these methods is that they need to fit a credible model to the plant I/O data. It is a non-trivial procedure [45], especially for closed-loop systems, involving data polishing, model structure selection, model fitting and validation. In addition, the reconfigured controller can still fail if the identified model is improperly post-processed, and its deployment may require an entire modification of the existing controller, which may also not be feasible in some systems.

In comparison with the foregoing model-based and data-driven methods, RL aided methods are competitive alternative approaches in dealing with performance optimization. First, they are data-driven methods that can be easily transplanted among different systems. Second, for performance optimization, they allow an optimal auxiliary controller to be learned straightforward through interactions with the existing system by trial and error with no need of system identification. The learned controller can also be placed into the system without the modification of the existing controller. Despite the benefits, little effort has been dedicated to the application of RL aided methods to performance optimization. The challenges lie in the following: 1) How to make RL aided methods work in a partially observable environment, where only a few plant states are accessible. This is very common in industrial systems, as it is costly to install sensors to observe all plant states. 2) How to make RL aided methods work in a noisy environment. Plant actuator and sensor noises are unavoidable, and the noises can render learning slow and unreliable. 3) How to make RL aided methods work with data-efficiency. Limited computational resources on board and limited time to recover system performance require this. 4) How to formulate performance indices related to system robustness and performance, so that RL aided methods can be applied to performance optimization.

All of these benefits and challenges motivate us to develop new practical methods for performance optimization of feedback control systems by harnessing the power of control theory and RL in the age of data.

1.2 Scope of the Work

For successful performance optimization of feedback control systems, two procedures can be performed: performance monitoring and controller reconfiguration. The structure is shown in Fig. 1.1. The performance monitoring procedure ascertains the cause, type and level of system performance degradation using plant I/O data through predefined performance indices [13, 14, 33, 36]. After a successful detection of performance degradation, a decision is made mainly according to the level of system degradation. To show the decision making visually, we assume that we have two performance indices J_1 and J_2 and define four regions of performance in Fig. 1.2. It can be seen from Fig. 1.2 that once the performance falls out of the region of required performance, a performance supervision system is activated attempting to bring the system performance back to the required one. This is where the controller reconfiguration procedure is performed. However, once the system performance enters the region of danger, a safety system should be invoked to interrupt the operation of the overall system to avoid danger for the system and its environment [9].

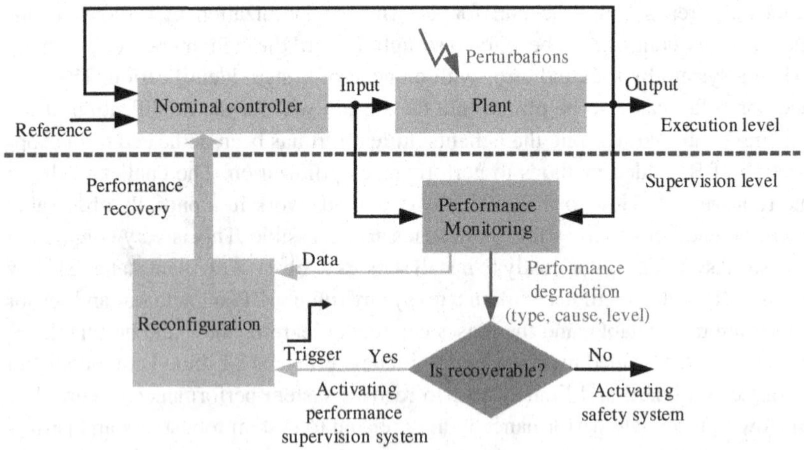

Figure 1.1 Structure for performance optimization

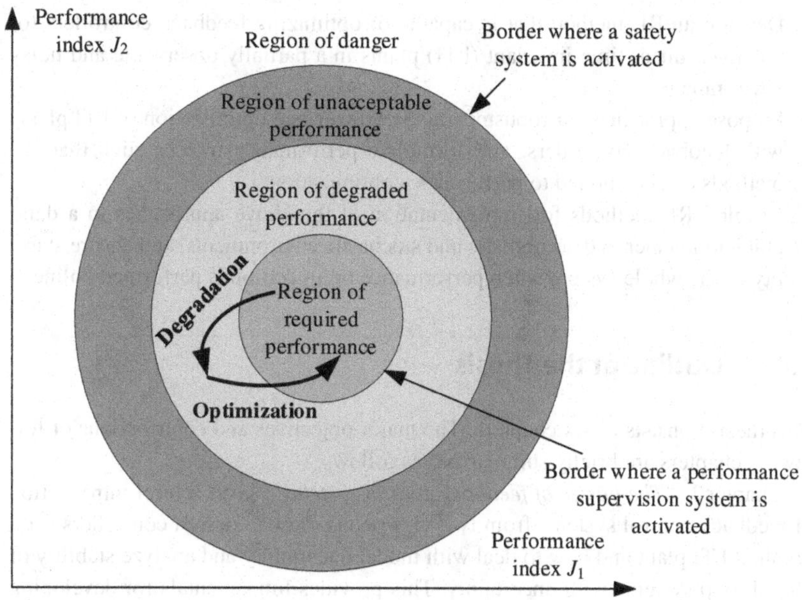

Figure 1.2 Regions of performance [9]

In industrial systems, the performance supervision system and the safety system work side-by-side with complementary aims. They represent two separate parts of control systems. For example, in the process industry, they are implemented in separate units. This separation allows the existing controller to be reconfigured without the need to meet safety standards [9].

In this work, we are only dedicated to developing new controller reconfiguration methods for the optimization of system performance in a data-driven manner. They are considered as a part of a performance supervision system, marked in green color in Fig. 1.1. The rest falls out of our scope.

1.3 Objective of the Work

The objective of the work is to develop RL aided methods for performance optimization of feedback control systems. To be specific, the main goals are stated as follows:

- Develop an RL method that is capable of optimizing feedback controllers for unknown linear time-invariant (LTI) plants in a partially observable and noisy environment.
- Propose approaches for robustness and performance optimization of LTI plants with feedback controllers, and formulate performance indices such that RL methods can be applied to performance optimization.
- Develop RL methods for implementation of the above approaches in a data-efficient manner in deterministic and stochastic environments, and ensure stability of the whole system when performance optimization is performed online.

1.4 Outline of the Thesis

This thesis consists of six chapters. The major objectives and contributions of following chapters are briefly summarized as follows.

Chapter 2, *"The basics of feedback control systems"*, gives a brief introduction of feedback control systems from two viewpoints: how to design controllers for a nominal LTI plant and how to deal with model uncertainty and analyze stability of closed-loop systems with uncertainty. This provides fundamentals for developing new performance optimization approaches.

Chapter 3, *"Reinforcement learning and feedback control"*, focuses on applying RL methods to solve two fundamental feedback control problems, linear quadratic regulator and linear quadratic Gaussian. These RL methods provide tools for data-driven implementation of new performance optimization approaches. This chapter can be read separately from Chapter 2. The key contribution is that we develop a new natural actor-critic (NAC) method that can learn an optimal output feedback controller with a prescribed structure for an LTI plant in a partially observable and noisy environment.

Chapter 4, "*Q-learning aided performance optimization of deterministic systems*", we propose two approaches, an input and output recovery approach, and a performance index based approach, for robustness and performance optimization of plants with observer-based state feedback controllers. We also generalize the latter approach to tracking performance optimization. Besides, we develop Q-learning methods for implementation of these approaches in a deterministic environment.

Chapter 5, *"NAC aided performance optimization of stochastic systems"*, focuses on the data-driven implementation of the above two performance optimization approaches in a stochastic environment using the NAC learning method. We also generalize the performance index based approach to optimize performance of plants with general output feedback controllers and demonstrate its effectiveness by a benchmark study on a BLDC motor test rig.

Finally, Chapter 6, "*Conclusion and future work*", gives a conclusion and discusses future work.

2 The Basics of Feedback Control Systems

In this chapter, an overview of feedback control systems is given. This consists of two parts. The first part concerns the design of feedback controllers for a nominal plant. The second part deals with the description of model uncertainty and the analysis of stability of closed-loop systems with uncertainty.

2.1 Design of Feedback Controllers

2.1.1 Description of a Nominal System

Let the minimal state-space representation of a nominal linear time-invariant (LTI) plant $\boldsymbol{P_0}$ be

$$\boldsymbol{P_0}: \begin{cases} \boldsymbol{x}(k+1) = \boldsymbol{A}\boldsymbol{x}(k) + \boldsymbol{B}\boldsymbol{u}(k), \quad \boldsymbol{x}(0) = \boldsymbol{x_0} \\ \boldsymbol{y}(k) = \boldsymbol{C}\boldsymbol{x}(k) + \boldsymbol{D}\boldsymbol{u}(k), \end{cases} \tag{2.1}$$

where $k \in \mathbb{Z}^+ = \{0, 1, 2, \cdots\}$ denotes sampling instants, $\boldsymbol{x} \in \mathbb{R}^n$, $\boldsymbol{u} \in \mathbb{R}^{k_u}$ and $\boldsymbol{y} \in \mathbb{R}^m$ are the plant state, input and output vectors, respectively. $\boldsymbol{x_0}$ is the initial plant state. $\boldsymbol{A}$, $\boldsymbol{B}$, $\boldsymbol{C}$ and $\boldsymbol{D}$ are real constant matrices with proper dimensions.

It is required that the pair $(\boldsymbol{A}, \boldsymbol{B})$ of $\boldsymbol{P_0}$ should be controllable or at least stabilizable, and the pair $(\boldsymbol{C}, \boldsymbol{A})$ should be observable or at least detectable, see their definitions in [85]. A lack of controllability or observability shows that more actuators or sensors are needed for effective control. Therefore, in our study we only deal with plants that are both controllable and observable.

The transfer function (matrix) of $\boldsymbol{P_0}$ can be represented by

$$\boldsymbol{P_0}(z) = \boldsymbol{C}(z\boldsymbol{I} - \boldsymbol{A})^{-1}\boldsymbol{B} + \boldsymbol{D}, \tag{2.2}$$

C. Hua, *Reinforcement Learning Aided Performance Optimization of Feedback Control Systems*, https://doi.org/10.1007/978-3-658-33034-7_2

where z denotes a complex variable of z-transform for a discrete-time signal. Here $\boldsymbol{P}_0(\infty) = \boldsymbol{D}$ and $\boldsymbol{P}_0$ is by definition a proper transfer matrix, namely, $\boldsymbol{P}_0 \in \mathcal{R}_p$. A special case is $\boldsymbol{D} = \boldsymbol{0}$, then we say $\boldsymbol{P}_0$ is a strictly proper transfer matrix, namely, $\boldsymbol{P}_0 \in \mathcal{R}_{sp}$. For brevity, later we use the following block partition notation to represent a transfer matrix

$$\left[\begin{array}{c|c} A & B \\ \hline C & D \end{array}\right] := C(zI - A)^{-1}B + D. \tag{2.3}$$

A stabilizing nominal observer-based state feedback controller $\boldsymbol{K}_0 \in \mathcal{R}_{sp}$ for $\boldsymbol{P}_0$ can be designed with a state feedback gain $\boldsymbol{F}$ and an observer gain $\boldsymbol{L}$ ensuring that $\boldsymbol{A} + \boldsymbol{BF}$ and $\boldsymbol{A} - \boldsymbol{LC}$ are Schur matrices, that is

$$\boldsymbol{K}_0 : \begin{cases} \hat{\boldsymbol{x}}(k+1) = \boldsymbol{A}\hat{\boldsymbol{x}}(k) + \boldsymbol{Bu}(k) + \boldsymbol{Lr}(k), \\ \boldsymbol{r}(k) = \boldsymbol{y}(k) - \boldsymbol{C}\hat{\boldsymbol{x}}(k) - \boldsymbol{Du}(k), \\ \boldsymbol{u}(k) = \boldsymbol{F}\hat{\boldsymbol{x}}(k), \end{cases} \tag{2.4}$$

where $\hat{\boldsymbol{x}}$ is the estimated state vector satisfying that $\forall \boldsymbol{x}(0)$, $\hat{\boldsymbol{x}}(0)$ and $\boldsymbol{u}(k)$, $\lim_{k\to\infty} \left(\boldsymbol{x}(k) - \hat{\boldsymbol{x}}(k)\right) = \boldsymbol{0}$, and $\boldsymbol{r}$ is the residual vector. Accordingly, $\boldsymbol{K}_0(z)$ can be denoted by

$$\boldsymbol{u}(z) = \boldsymbol{K}_0(z)\,\boldsymbol{y}(z)$$

and

$$\boldsymbol{K}_0(z) = \left[\begin{array}{c|c} \boldsymbol{A} + \boldsymbol{BF} - \boldsymbol{LC} - \boldsymbol{LDF} & \boldsymbol{L} \\ \hline \boldsymbol{F} & \boldsymbol{0} \end{array}\right].$$

It can be checked that the poles of the nominal system, consisting of the pair $(\boldsymbol{P}_0, \boldsymbol{K}_0)$, are determined, exclusively, by the eigenvalues of $\boldsymbol{A} + \boldsymbol{BF}$ and $\boldsymbol{A} - \boldsymbol{LC}$. As $\boldsymbol{P}_0$ is controllable and observable, the design of $\boldsymbol{F}$ and $\boldsymbol{L}$ enables the system poles to be arbitrarily assigned to achieve desired system performance. We will investigate performance evaluation and optimization of a controller in Chapter 3.

2.1.2 A Coprime Factorization Design Tool

For analysis of plants and feedback controllers, the coprime factorization of transfer matrices is introduced. It says that a transfer matrix can be factorized into the product of two transfer matrices, one is stable and the other has a stable inverse. Coprimeness is expressed as a full rank condition on the two transfer matrices in $|z| > 1$ [68].

Definition 2.1 *Two transfer matrices $\hat{M}(z)$ and $\hat{N}(z)$ are left coprime over $\mathcal{RH}_\infty$, if $[\hat{M}(z)\ \hat{N}(z)]$ has full row rank in $|z| > 1$, or equivalently, if there exist two transfer matrices $\hat{X}(z)$ and $\hat{Y}(z)$ in $\mathcal{RH}_\infty$ such that*

$$\left[\hat{M}(z)\ \hat{N}(z)\right]\begin{bmatrix}\hat{X}(z)\\ \hat{Y}(z)\end{bmatrix} = I.$$

Similarly, two transfer matrices $M(z)$ and $N(z)$ are right coprime over $\mathcal{RH}_\infty$, if $[M^T(z)\ N^T(z)]^T$ has full column rank in $|z| > 1$, or equivalently, if there exist two transfer matrices $X(z)$ and $Y(z)$ in $\mathcal{RH}_\infty$ such that

$$\left[X(z)\ Y(z)\right]\begin{bmatrix}M(z)\\ N(z)\end{bmatrix} = I.$$

Definition 2.2 *$P_0(z) = \hat{M}^{-1}(z)\hat{N}(z)$ with $\hat{M}(z)$ and $\hat{N}(z)$ left coprime over $\mathcal{RH}_\infty$ is called a left coprime factorization (LCF) of $P_0(z)$. Similarly, $P_0(z) = N(z)M^{-1}(z)$ with $M(z)$ and $N(z)$ right coprime over $\mathcal{RH}_\infty$ is called a right coprime factorization (RCF) of P_0.*

The following lemma [13] shows the state-space computation of the LCF and RCF of both the nominal plant P_0 and the nominal controller K_0, which serves as a cornerstone of our subsequent studies.

Lemma 2.1 *Consider the nominal plant P_0 (2.1) and the stabilizing nominal observer-based state feedback controller K_0 (2.4), and define*

$$\begin{bmatrix} X(z) & Y(z)\\ -\hat{N}(z) & \hat{M}(z)\end{bmatrix} = \left[\begin{array}{c|cc} A-LC & -(B-LD) & -L\\ \hline F & I & 0\\ C & -D & I\end{array}\right], \tag{2.5}$$

$$\begin{bmatrix} M(z) & -\hat{Y}(z)\\ N(z) & \hat{X}(z)\end{bmatrix} = \left[\begin{array}{c|cc} A+BF & B & L\\ \hline F & I & 0\\ C+DF & D & I\end{array}\right]. \tag{2.6}$$

Then

$$P_0(z) = \hat{M}^{-1}(z)\,\hat{N}(z) = N(z)\,M^{-1}(z), \tag{2.7}$$

$$K_0(z) = -X^{-1}(z)\,Y(z) = -\hat{Y}(z)\,\hat{X}^{-1}(z) \tag{2.8}$$

are the LCF and RCF of $\boldsymbol{P}_0(z)$, and those of $\boldsymbol{K}_0(z)$, respectively. Moreover, the Bézout identify holds

$$\begin{bmatrix} \boldsymbol{X}(z) & \boldsymbol{Y}(z) \\ -\hat{\boldsymbol{N}}(z) & \hat{\boldsymbol{M}}(z) \end{bmatrix} \begin{bmatrix} \boldsymbol{M}(z) & -\hat{\boldsymbol{Y}}(z) \\ \boldsymbol{N}(z) & \hat{\boldsymbol{X}}(z) \end{bmatrix} = \begin{bmatrix} \boldsymbol{I} & \boldsymbol{0} \\ \boldsymbol{0} & \boldsymbol{I} \end{bmatrix}. \tag{2.9}$$

2.1.3 Well-posedness and Internal Stability

To design a feedback controller to achieve desired performance, two conditions are required that are the physical realizability and stability of the system.

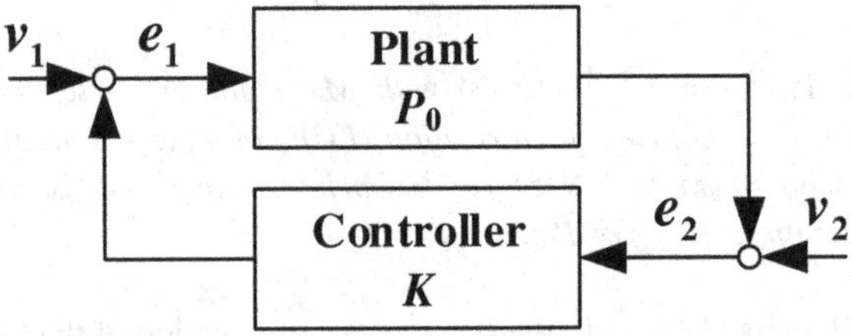

Figure 2.1 Internal stability analysis diagram

We study the two conditions with the feedback loop comprised of $\boldsymbol{P}_0$ (2.1) and a general feedback controller $\boldsymbol{K} \in \mathcal{R}_p$. It is shown in Fig. 2.1, where $\boldsymbol{e}_1$ and $\boldsymbol{e}_2$ are the input signals of $\boldsymbol{P}_0$ and $\boldsymbol{K}$, and $\boldsymbol{v}_1$ and $\boldsymbol{v}_2$ are external input signals.

First, the realizability condition is assured by the well-posedness property [85] that is defined below.

Definition 2.3 *A feedback system is said to be well-posed if all closed-loop transfer matrices are well-defined and proper.*

Lemma 2.2 *The feedback system in Fig. 2.1 is well-posed if and only if*

$$\boldsymbol{I} - \boldsymbol{K}(\infty)\, \boldsymbol{P}_0(\infty) \tag{2.10}$$

is invertible.

The well-posedness property can also be expressed in terms of state-space realizations. Assume $\boldsymbol{K}(z)$ has the realization:

$$\boldsymbol{K}(z) = \left[\begin{array}{c|c} \boldsymbol{A_K} & \boldsymbol{B_K} \\ \hline \boldsymbol{C_K} & \boldsymbol{D_K} \end{array}\right].$$

It is known that $\boldsymbol{P_0}(\infty) = \boldsymbol{D}$ and $\boldsymbol{K}(\infty) = \boldsymbol{D_K}$. Thus, the condition given in (2.10) is equivalent to that $\boldsymbol{I} - \boldsymbol{D_K}\boldsymbol{D}$ is invertible. Fortunately, most industrial systems have the well-posedness property, since in most plants we have $\boldsymbol{D} = \boldsymbol{0}$.

Next, considering that the feedback system in Fig. 2.1 is well-posed, the system stability condition is guaranteed by the internal stability property, which is defined by [85]:

Definition 2.4 *The feedback system in Fig. 2.1 is said to be internally stable if the closed-loop transfer matrix*

$$\begin{bmatrix} \boldsymbol{I} & -\boldsymbol{K} \\ -\boldsymbol{P_0} & \boldsymbol{I} \end{bmatrix}^{-1} = \begin{bmatrix} (\boldsymbol{I} - \boldsymbol{K}\boldsymbol{P_0})^{-1} & \boldsymbol{K}(\boldsymbol{I} - \boldsymbol{P_0}\boldsymbol{K})^{-1} \\ \boldsymbol{P_0}(\boldsymbol{I} - \boldsymbol{K}\boldsymbol{P_0})^{-1} & (\boldsymbol{I} - \boldsymbol{P_0}\boldsymbol{K})^{-1} \end{bmatrix} \tag{2.11}$$

from $\begin{bmatrix} \boldsymbol{v}_1 \\ \boldsymbol{v}_2 \end{bmatrix}$ *to* $\begin{bmatrix} \boldsymbol{e}_1 \\ \boldsymbol{e}_2 \end{bmatrix}$ *belongs to* $\mathcal{RH}_\infty$.

This basically says that all signals in a system are bounded provided that the injected signals (at any locations) are bounded. The internal stability property can also be formulated in terms of coprime factorizations [85]. Consider that $\boldsymbol{P_0}(z)$ and $\boldsymbol{K}(z)$ have the following LCF and RCF:

$$\begin{aligned} \boldsymbol{P_0}(z) &= \hat{\boldsymbol{M}}^{-1}(z)\,\hat{\boldsymbol{N}}(z) = \boldsymbol{N}(z)\,\boldsymbol{M}^{-1}(z) \\ \boldsymbol{K}(z) &= -\boldsymbol{X}^{-1}(z)\,\boldsymbol{Y}(z) = -\hat{\boldsymbol{Y}}(z)\,\hat{\boldsymbol{X}}^{-1}(z). \end{aligned} \tag{2.12}$$

Lemma 2.3 *Consider the feedback system in Fig. 2.1. The following conditions are equivalent:*

1. $(\boldsymbol{P_0}(z), \boldsymbol{K}(z))$ *is a stabilizing pair.*
2. $\begin{bmatrix} \boldsymbol{X}(z) & \boldsymbol{Y}(z) \\ -\hat{\boldsymbol{N}}(z) & \hat{\boldsymbol{M}}(z) \end{bmatrix}^{-1} \in \mathcal{RH}_\infty.$
3. $\begin{bmatrix} \boldsymbol{M}(z) & -\hat{\boldsymbol{Y}}(z) \\ \boldsymbol{N}(z) & \hat{\boldsymbol{X}}(z) \end{bmatrix}^{-1} \in \mathcal{RH}_\infty.$

We extend the internal stability result to the design of feedforward/feedback (also called two-degree-of-freedom) controller for the plant $\boldsymbol{P}_0$. Consider the system shown in Fig. 2.2, where $\boldsymbol{d}$ is a reference input vector, and $\boldsymbol{V} \in \mathcal{R}_p$ is a feedforward controller. Both $\boldsymbol{d}$ and $\boldsymbol{y}$ are used to generate the control input $\boldsymbol{u}$. For stability analysis, an equivalent description of the system with an augmented plant $\boldsymbol{P}_A \in \mathcal{R}_p$ and a feedback controller $\boldsymbol{K}_A \in \mathcal{R}_p$ is shown in Fig. 2.3. In order to guarantee its internal stability, according to (2.11), the following condition should be satisfied [68]:

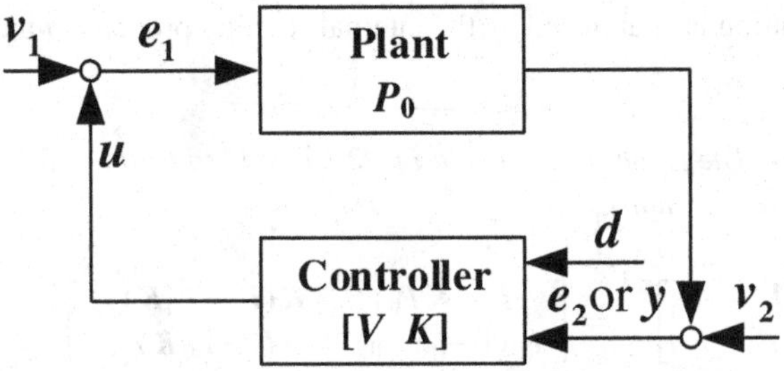

Figure 2.2 Representation of a feedforward/feedback controller

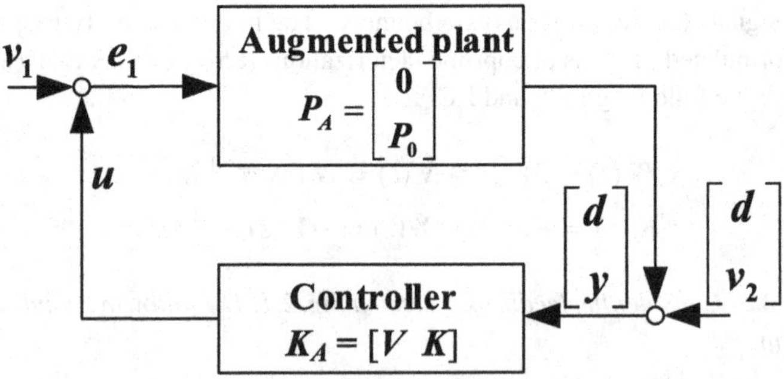

Figure 2.3 Representation of a feedforward/feedback controller as a feedback controller for an augmented plant

$$\begin{bmatrix} I & -K_A \\ -P_A & I \end{bmatrix}^{-1} = \begin{bmatrix} I & -\begin{bmatrix} V & K \end{bmatrix} \\ -\begin{bmatrix} 0 \\ P_0 \end{bmatrix} & \begin{bmatrix} I & 0 \\ 0 & I \end{bmatrix} \end{bmatrix}^{-1}$$

$$= \begin{bmatrix} (I - KP_0)^{-1} & (I - KP_0)^{-1}V & K(I - P_0K)^{-1} \\ 0 & I & 0 \\ P_0(I - KP_0)^{-1} & P_0(I - KP_0)^{-1}V & (I - P_0K)^{-1} \end{bmatrix} \in \mathcal{RH}_\infty. \quad (2.13)$$

Based on (2.13), the internal stability condition can be interpreted as [68]:

- If V and K are designed separately, then $V \in \mathcal{RH}_\infty$ should hold and K should stabilize P_0.
- If V and K are designed together as one controller K_A, then K should stabilize P_0, and any unstable modes in V should be included in K. From the viewpoint of coprime factoriztion, this condition amounts to that if P_0 and K have coprime factoriztions (2.12), then Lemma 2.3 holds and V has a coprime factorization

$$V(z) = -X^{-1}(z)\, Y_d(z), \quad (2.14)$$

where $Y_d(z) \in \mathcal{RH}_\infty$ holds.

2.1.4 Parameterization of Stabilizing Controllers

Crucial to our study is the characterization of the class of all stabilizing controllers for P_0 in terms of a parameter termed Q_r. The parameterization has been developed in a continuous-time setting by Youla, Bongiorno and Jabr [80, 81], and in a discrete-time setting by Kučera [40].

Theorem 2.1 *Given the plant P_0 (2.1) and the nominal controller K_0 (2.4) with coprime factorizations (2.7), (2.8), then the class of all stabilizing feedback controllers for P_0 can be left and right coprime parameterized by*

$$K(Q_r) = -(X + Q_r\hat{N})^{-1}(Y - Q_r\hat{M}) = -(\hat{Y} - MQ_r)(\hat{X} + NQ_r)^{-1}, \quad (2.15)$$

where $Q_r \in \mathcal{RH}_\infty$ is called Youla-Kučera (YK) parameter, and trivially, $K(\mathbf{0}) = K_0$.

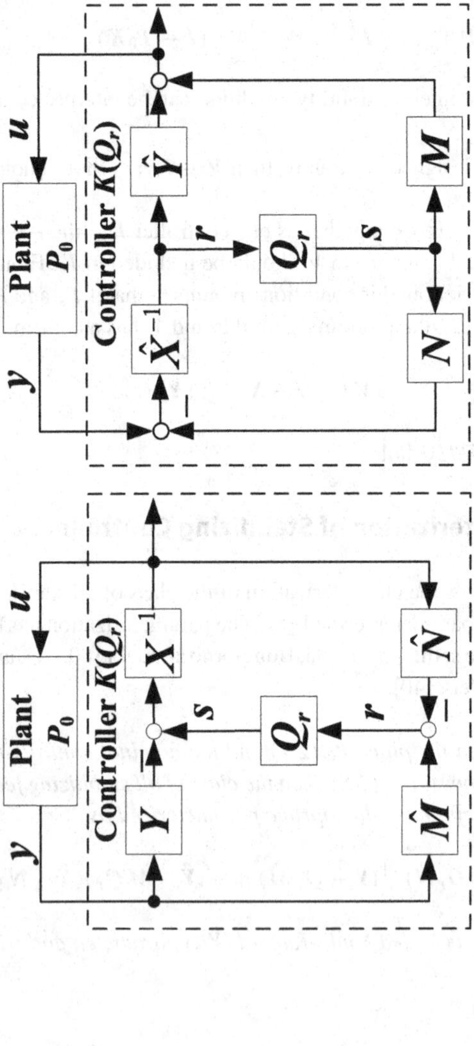

Figure 2.4 LCF and RCF of $K(Q_r)$

The LCF and RCF of $\boldsymbol{K}(\boldsymbol{Q}_r)$ are shown in Fig. 2.4, where $\boldsymbol{s}$ is the output vector of $\boldsymbol{Q}_r$ that is a compensation signal, and $\boldsymbol{r}$ is the residual vector. The relationship between $\boldsymbol{u}$, $\boldsymbol{y}$ and $\boldsymbol{s}$, $\boldsymbol{r}$ can be represented by [6, 68]

$$\begin{bmatrix} u \\ y \end{bmatrix} = \begin{bmatrix} M & -\hat{Y} \\ N & \hat{X} \end{bmatrix} \begin{bmatrix} s \\ r \end{bmatrix}, \begin{bmatrix} s \\ r \end{bmatrix} = \begin{bmatrix} X & Y \\ -\hat{N} & \hat{M} \end{bmatrix} \begin{bmatrix} u \\ y \end{bmatrix}. \tag{2.16}$$

The LCF of $\boldsymbol{K}(\boldsymbol{Q}_r)$ (2.15) can also be, equivalently, realized in an observer-based form [14, 85]

$$\boldsymbol{K}(\boldsymbol{Q}_r): \begin{cases} \hat{\boldsymbol{x}}(k+1) = \boldsymbol{A}\hat{\boldsymbol{x}}(k) + \boldsymbol{B}\boldsymbol{u}(k) + \boldsymbol{L}\boldsymbol{r}(k), \\ \boldsymbol{r}(k) = -\boldsymbol{C}\hat{\boldsymbol{x}}(k) - \boldsymbol{D}\boldsymbol{u}(k) + \boldsymbol{y}(k), \\ \boldsymbol{u}(k) = \boldsymbol{F}\hat{\boldsymbol{x}}(k) + \boldsymbol{s}(k), \\ \boldsymbol{s}(z) = \boldsymbol{Q}_r(z)\boldsymbol{r}(z). \end{cases} \tag{2.17}$$

It is shown in Fig. 2.5. Next, we parameterize all stabilizing two-degree-of-freedom (2-DOF) controllers for $\boldsymbol{P}_0$. Considering a nominal controller $\boldsymbol{K}_A = [\boldsymbol{V}\ \boldsymbol{K}_0]$ that stabilizes the augmented plant $\boldsymbol{P}_A$ with the structure shown in Fig. 2.3, and $\boldsymbol{P}_0$, $\boldsymbol{K}_0$, $\boldsymbol{V}$ have coprime factorizations (2.7), (2.8) and (2.14), then $\boldsymbol{K}_A$ and $\boldsymbol{P}_A$ can be coprime factorized by

$$\boldsymbol{K}_A = -\boldsymbol{X}_A^{-1}\boldsymbol{Y}_A, \ \boldsymbol{P}_A = \hat{\boldsymbol{M}}_A^{-1}\hat{\boldsymbol{N}}_A,$$

$$\boldsymbol{X}_A = \boldsymbol{X}, \boldsymbol{Y}_A = [\boldsymbol{Y}_d\ \boldsymbol{Y}], \hat{\boldsymbol{M}}_A = \begin{bmatrix} \boldsymbol{I} & 0 \\ 0 & \hat{\boldsymbol{M}} \end{bmatrix}, \hat{\boldsymbol{N}}_A = \begin{bmatrix} 0 \\ \hat{\boldsymbol{N}} \end{bmatrix},$$

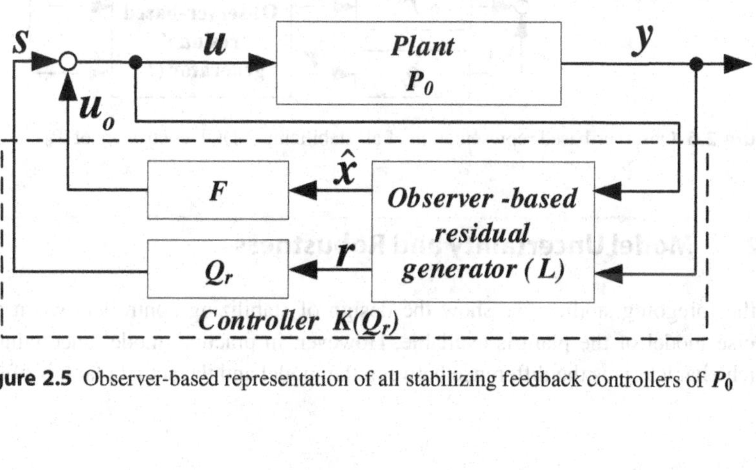

Figure 2.5 Observer-based representation of all stabilizing feedback controllers of $\boldsymbol{P}_0$

and all 2-DOF stabilizing controllers for P_0, according to Theorem 2.1, can be parameterized by

$$K_A\left(\bar{Q}_r\right) = -\left(X_A + \bar{Q}_r\hat{N}_A\right)^{-1}\left(Y_A - \bar{Q}_r\hat{M}_A\right),$$

where $\bar{Q}_r$ belongs to $\mathcal{RH}_\infty$. Accordingly, they can be represented in the following state-space form

$$\begin{cases} \hat{x}(k+1) = A\hat{x}(k) + Bu(k) + Lr(k), \\ r(k) = -C\hat{x}(k) - Du(k) + y(k), \\ u(k) = F\hat{x}(k) + s(k) + h(k), \\ s(z) = Q_r(z)r(z), \\ h(z) = (-Y_d(z) + Q_d(z))\,d(z), \end{cases} \tag{2.18}$$

where $\bar{Q}_r = [Q_d \;\; Q_r]$, and $h(k)$ denotes the feedforward of the reference input $d(k)$ to the plant input. The whole control diagram is described in Fig. 2.6. It is worth mentioning that Q_d and Q_r are two design parameters useful for tracking and robustness performance optimization, since the system stability is still preserved as long as they are stable. Readers are referred to [19, 21, 24, 47] for their applications.

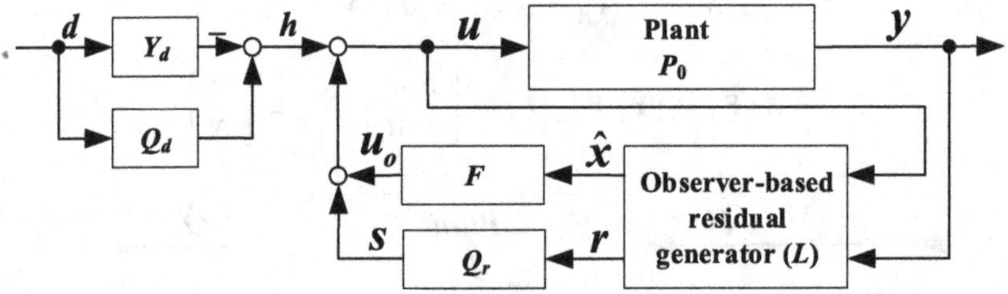

Figure 2.6 Observer-based representation of all stabilizing 2-DOF controllers of P_0

2.2 Model Uncertainty and Robustness

In the foregoing studies, we show the design of stabilizing controllers when an precise model of the plant is available. However, in practice, model uncertainty, which characterizes the difference between the model and the actual plant, is often inevitable. In this section, we deal with descriptions of the plant uncertainty, and give conditions for the design of feedback controllers to maintain overall stability despite the uncertainty.

2.2.1 Small Gain Theorem

The small gain theorem [12, 85] provides an essential tool for analysis of the closed-loop stability. Consider a feedback system shown in Fig. 2.7 with the plant regarded as an unstructured uncertainty $\boldsymbol{\Delta}(z)$ and a controller $\boldsymbol{K}(z)$.

Theorem 2.2 *Suppose* $\boldsymbol{K}(z) \in \mathcal{RH}_\infty$ *and let* $\varepsilon > 0$*. Then the feedback system shown in Fig. 2.7 is well-posed and internally stable,* $\forall \boldsymbol{\Delta}(z) \in \mathcal{RH}_\infty$*, with*

(a) $\|\boldsymbol{\Delta}(z)\|_\infty \leq 1/\varepsilon$ *if and only if* $\|\boldsymbol{K}(z)\|_\infty < \varepsilon$*.*
(b) $\|\boldsymbol{\Delta}(z)\|_\infty < 1/\varepsilon$ *if and only if* $\|\boldsymbol{K}(z)\|_\infty \leq \varepsilon$*.*

Remark 2.1 *The small gain theorem is a sufficient but not a necessary condition for a fixed type of uncertainty. Here is a simple example. If* $\boldsymbol{\Delta}(z)$ *and* $\boldsymbol{K}(z)$ *are fixed scalar transfer functions, Theorem 2.2 requires the closed-loop gain satisfying* $|\boldsymbol{K}(e^{j\theta})\boldsymbol{\Delta}(e^{j\theta})| < 1, \forall\theta \in [0, 2\pi)$ *to ensure stability. This corresponds to a positive gain margin and an infinite phase margin according to the Nyquist criterion [52].*

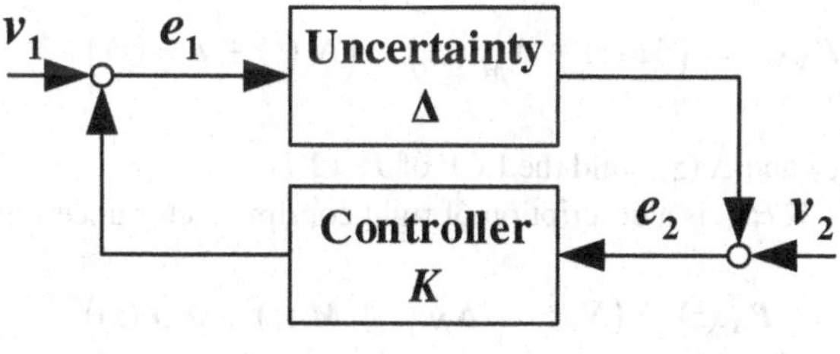

Figure 2.7 System representation for robust stability analysis

The small gain theorem cannot be applied directly to controller synthesis in occasions where a high closed-loop gain is required for good performance, e.g. design of a PID controller for tracking performance. However, it can be useful for synthesizing controllers for plants with uncertainty to improve stability margins, see [43] for in-depth treatment.

2.2.2 Coprime Factor Representation of Model Uncertainty

There are two broad types of uncertainty descriptions: structured and unstructured uncertainty. Structured uncertainty often associates uncertainty with the change of concrete physical parameters, such as the change of resistance or capacitance in electrical components. In this thesis, we mainly discuss descriptions of unstructured uncertainty, as their stability results can be easily extended to the structured case.

The unstructured uncertainty is commonly described in an additive, multiplicative or coprime factorized manner, see [11, 16, 72, 73]. Let $\boldsymbol{P}_0(z)$ and $\boldsymbol{P}_\Delta(z)$ be the transfer matrices of nominal and uncertain plants, then

1. $\boldsymbol{\Delta}(z) \in \mathcal{RH}_\infty$ is a description of additive uncertainty if
$$\boldsymbol{P}_\Delta(z) = \boldsymbol{P}_0(z) + \boldsymbol{\Delta}(z).$$
2. $\boldsymbol{\Delta}(z) \in \mathcal{RH}_\infty$ is a description of multiplicative uncertainty if
$$\boldsymbol{P}_\Delta(z) = \boldsymbol{P}_0(z)\big(\boldsymbol{I} + \boldsymbol{\Delta}(z)\big).$$
3. $\boldsymbol{\Delta}_{\hat{M}}, \boldsymbol{\Delta}_{\hat{N}} \in \mathcal{RH}_\infty$ is a description of left coprime factor uncertainty if
$$\boldsymbol{P}_\Delta(z) = \left(\hat{\boldsymbol{M}}(z) + \boldsymbol{\Delta}_{\hat{M}}(z)\right)^{-1}\left(\hat{\boldsymbol{N}}(z) + \boldsymbol{\Delta}_{\hat{N}}(z)\right), \tag{2.19}$$
where $\hat{\boldsymbol{M}}(z)$ and $\hat{\boldsymbol{N}}(z)$ build the LCF of $\boldsymbol{P}_0$ (2.7).
4. $\boldsymbol{\Delta}_M, \boldsymbol{\Delta}_N \in \mathcal{RH}_\infty$ is a description of right coprime factor uncertainty if
$$\boldsymbol{P}_\Delta(z) = \big(\boldsymbol{N}(z) + \boldsymbol{\Delta}_N(z)\big)\big(\boldsymbol{M}(z) + \boldsymbol{\Delta}_M(z)\big)^{-1}, \tag{2.20}$$
where $\boldsymbol{M}(z)$ and $\boldsymbol{N}(z)$ build the RCF of $\boldsymbol{P}_0$ (2.7).

It has been pointed out in [21, 73] that coprime factors are powerful in characterizing plant uncertainty as they capture both stable and unstable pole and zero uncertainty. Thus, in what follows, we analyze the feedback system stability by considering the coprime factor uncertainty of the plant.

Consider the closed-loop consisting of $P_\Delta(z)$ described with $P_0(z)$ and the above left coprime factor uncertainty, and a stabilizing controller $K(Q_r)$ (2.15) for $P_0(z)$ in Fig. 2.8. Then the closed-loop stability can be analyzed by the following theorem:

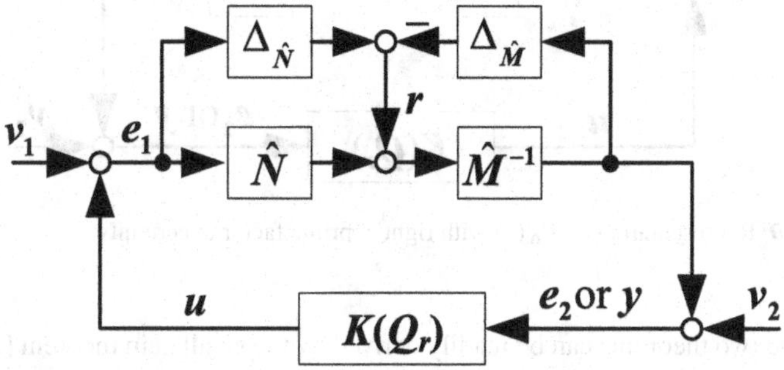

Figure 2.8 Representation of $P_\Delta(z)$ with left coprime factor uncertainty

Theorem 2.3 *Let the uncertain plant $P_\Delta(z)$ (2.19) be described with $P_0(z)$ and left coprime factor uncertainty, and $K(Q_r)$ (2.15) be a stabilizing controller for the nominal plant $P_0(z)$ and $\varepsilon > 0$, then the following statements are equivalent:*

(a) The closed-loop system consisting of $P_\Delta(z)$ and $K(Q_r)$ shown in Fig. 2.8 is internally stable, $\forall P_\Delta(z)$ with $\|[\Delta_{\hat{M}}\ \Delta_{\hat{N}}]\|_\infty \le \varepsilon$.

(b) $\left\| \begin{bmatrix} -\hat{Y} \\ \hat{X} \end{bmatrix} + \begin{bmatrix} M \\ N \end{bmatrix} Q_r \right\|_\infty < 1/\varepsilon.$

Alternatively, considering right coprime factor uncertainty shown in Fig. 2.9, the following theorem holds:

Theorem 2.4 *Let the uncertain plant $P_\Delta(z)$ (2.20) be described with $P_0(z)$ and right coprime factor uncertainty, and $K(Q_r)$ (2.15) be a stabilizing controller for the nominal plant $P_0(z)$ and $\varepsilon > 0$, then the following statements are equivalent:*

(a) The closed-loop system consisting of $P_\Delta(z)$ and $K(Q_r)$ shown in Fig. 2.9 is internally stable, $\forall P_\Delta(z)$ with $\left\| \begin{bmatrix} \Delta_M \\ \Delta_N \end{bmatrix} \right\|_\infty \le \varepsilon$.

(b) $\left\| \begin{bmatrix} Y & X \end{bmatrix} + Q_r \begin{bmatrix} \hat{M} & -\hat{N} \end{bmatrix} \right\|_\infty < 1/\varepsilon.$

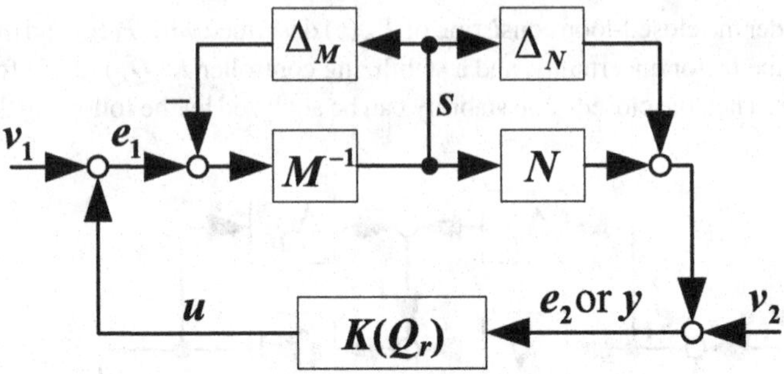

Figure 2.9 Representation of $P_\Delta(z)$ with right coprime factor uncertainty

The above two theorems can be readily derived by the small gain theorem [22, 46]. They will be later used as a conservative condition for the examination of closed-loop stability.

2.2.3 Dual YK Representation of Model Uncertainty

Unlike the coprime factor uncertainty, this part presents a new representation of model uncertainty that is useful for controller synthesis.

Dual to the characterization of the class of all stabilizing controllers for a nominal plant by a stable transfer matrix given in Theorem 2.1, the class of all plants stabilized by a given controller can be characterized by a stable transfer matrix. Consider the nominal plant $P_0(z)$ (2.1) and the stabilizing nominal controller $K_0(z)$ (2.4), all plants stabilized by $K_0(z)$ are given by the following dual YK parameterization theorem [68]:

Theorem 2.5 *For the coprime factorizations of* P_0 *(2.7) and* K_0 *(2.8), the class of uncertain plants stabilized by* K_0 *can be left and right coprime parameterized by*

$$P(S_f) = (\hat{M} - S_f Y)^{-1}(\hat{N} + S_f X) = (N + \hat{X} S_f)(M - \hat{Y} S_f)^{-1}, \qquad (2.21)$$

where $S_f \in \mathcal{RH}_\infty$ *is called dual YK parameter that represents the uncertainty, and trivially,* $P(0) = P_0$.

The interconnection between the uncertainty $\boldsymbol{S}_f$ and the coprime factor uncertainty introduced in the last subsection can be found in [51]. According to the above theorem, if both the actual plant $\boldsymbol{P}(\boldsymbol{S}_f)$ and the nominal plant $\boldsymbol{P}_0$ are stabilized by $\boldsymbol{K}_0$, then the uncertainty $\boldsymbol{S}_f$ can be determined, considering the Bézout identity (2.9), by [68]

$$\boldsymbol{S}_f = \hat{\boldsymbol{X}}^{-1}\left(\boldsymbol{I} - \boldsymbol{P}\left(\boldsymbol{S}_f\right)\boldsymbol{K}_0\right)^{-1}\left(\boldsymbol{P}\left(\boldsymbol{S}_f\right) - \boldsymbol{P}_0\right)\boldsymbol{M}. \tag{2.22}$$

Furthermore, $\boldsymbol{S}_f$ represents the transfer matrix from the signal $\boldsymbol{s}$ to $\boldsymbol{r}$ seen in Fig. 2.5. This can be derived by describing the RCF of $\boldsymbol{P}(\boldsymbol{S}_f)$ in an observer-based form:

$$\begin{cases} \hat{\boldsymbol{x}}(k+1) = \boldsymbol{A}\hat{\boldsymbol{x}}(k) + \boldsymbol{B}\boldsymbol{u}(k) + \boldsymbol{L}\boldsymbol{r}(k), \\ \quad \boldsymbol{u}(k) = \boldsymbol{F}\hat{\boldsymbol{x}}(k) + \boldsymbol{s}(k), \\ \quad \boldsymbol{y}(k) = \boldsymbol{C}\hat{\boldsymbol{x}}(k) + \boldsymbol{D}\boldsymbol{u}(k) + \boldsymbol{r}(k), \\ \quad \boldsymbol{r}(z) = \boldsymbol{S}_f(z)\boldsymbol{s}(z), \\ \Rightarrow \boldsymbol{u}(z) = \left(\boldsymbol{M}(z) - \hat{\boldsymbol{Y}}(z)\boldsymbol{S}_f(z)\right)\boldsymbol{s}(z), \ \boldsymbol{y}(z) = \left(\boldsymbol{N}(z) + \hat{\boldsymbol{X}}(z)\boldsymbol{S}_f(z)\right)\boldsymbol{s}(z). \end{cases} \tag{2.23}$$

Note that $\boldsymbol{S}_f$ can be useful to represent system performance degradation induced by various factors, such as additive disturbances, parametric and structural system changes. Subsequently, we consider the class of plants $\boldsymbol{P}(\boldsymbol{S}_f)$ stabilized by $\boldsymbol{K}_0$ and the class of controllers $\boldsymbol{K}(\boldsymbol{Q}_r)$ (2.15) stabilized by $\boldsymbol{P}_0$. The stability of the closed-loop system formed by $\boldsymbol{P}(\boldsymbol{S}_f)$ and $\boldsymbol{K}(\boldsymbol{Q}_r)$ is given by the following theorem [68]:

Theorem 2.6 *Let* $(\boldsymbol{P}_0, \boldsymbol{K}_0)$ *be a stabilizing plant controller pair. They have coprime factorizations (2.7) and (2.8). Consider* $\boldsymbol{P}(\boldsymbol{S}_f)$ *(2.21) and* $\boldsymbol{K}(\boldsymbol{Q}_r)$ *(2.15) with* $\boldsymbol{S}_f, \boldsymbol{Q}_r \in \mathcal{RH}_\infty$. *Then the pair* $\left(\boldsymbol{P}(\boldsymbol{S}_f), \boldsymbol{K}(\boldsymbol{Q}_r)\right)$ *is stabilizing if and only if the pair* $\left(\boldsymbol{S}_f, \boldsymbol{Q}_r\right)$ *is stabilizing as shown in Fig. 2.10. In particular:*

$$\begin{aligned} &\begin{bmatrix} \boldsymbol{I} & -\boldsymbol{K}(\boldsymbol{Q}_r) \\ -\boldsymbol{P}(\boldsymbol{S}_f) & \boldsymbol{I} \end{bmatrix}^{-1} \\ &= \begin{bmatrix} \boldsymbol{I} & -\boldsymbol{K}_0 \\ -\boldsymbol{P}_0 & \boldsymbol{I} \end{bmatrix}^{-1} + \begin{bmatrix} \boldsymbol{M} & -\hat{\boldsymbol{Y}} \\ \boldsymbol{N} & \hat{\boldsymbol{X}} \end{bmatrix} \left\{ \begin{bmatrix} \boldsymbol{I} & -\boldsymbol{Q}_r \\ -\boldsymbol{S}_f & \boldsymbol{I} \end{bmatrix}^{-1} - \boldsymbol{I} \right\} \begin{bmatrix} \boldsymbol{X} & \boldsymbol{Y} \\ -\hat{\boldsymbol{N}} & \hat{\boldsymbol{M}} \end{bmatrix}. \end{aligned} \tag{2.24}$$

It can be concluded from Theorem 2.6 that the performance of the system consisting of $\boldsymbol{P}(\boldsymbol{S}_f)$ and $\boldsymbol{K}(\boldsymbol{Q}_r)$ can be adjusted by tuning $\boldsymbol{Q}_r$ without spoiling the system stability as long as $\boldsymbol{Q}_r$ stabilizes $\boldsymbol{S}_f$. A common practice for the design of $\boldsymbol{Q}_r$ to improve system performance is pointed out in [3, 68], including two stages:

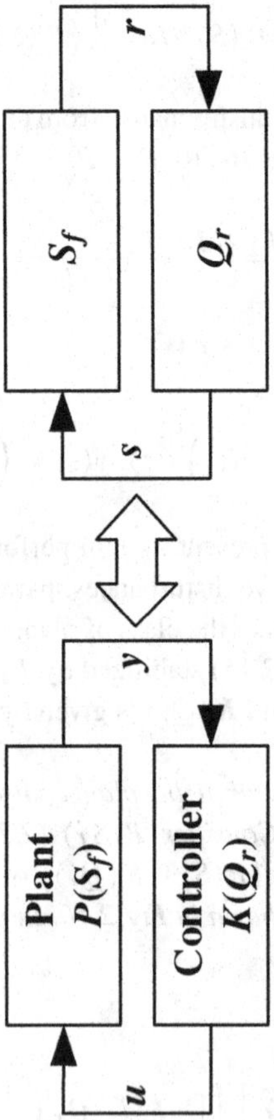

Figure 2.10 Equivalent robust stability property

- the identification of the uncertainty $\boldsymbol{S}_f$ using measurable data $\boldsymbol{s}$ and $\boldsymbol{r}$ [29, 30, 69];
- the design of $\boldsymbol{Q}_r$ based on the identified $\boldsymbol{S}_f$ using design techniques, such as pole placement, linear quadratic Gaussian or $\mathcal{H}_\infty$ [69, 79].

However, there are several issues regarding the design of $\boldsymbol{Q}_r$: 1) It is hard to fit a credible model of $\boldsymbol{S}_f$ to the measurable data. According to Equation (2.22), the actual order of $\boldsymbol{S}_f$ could be as high as the sum of the order of $\boldsymbol{P}(\boldsymbol{S}_f)$ and $\boldsymbol{P}_0$. 2) An improper post-processing of the identified $\boldsymbol{S}_f$ can lead to the failure of the design of $\boldsymbol{Q}_r$. 3) It is hard to quantify the loss of optimality of a design achieved by the two-step procedure comprised of the design of $\boldsymbol{K}_0$ for $\boldsymbol{P}_0$ and the design of $\boldsymbol{Q}_r$ for $\boldsymbol{S}_f$ in contrast to an optimal design of $\boldsymbol{K}(\boldsymbol{Q}_r)$ for $\boldsymbol{P}(\boldsymbol{S}_f)$.

To get around the foregoing issues, unlike these two-stage design approaches, we will deal with the design of $\boldsymbol{Q}_r$ and other useful control structures for performance optimization of feedback systems using reinforcement learning methods without the identification of the uncertainty $\boldsymbol{S}_f$. They will be introduced in Chapter 4 and Chapter 5.

2.3 Concluding Remarks

This chapter gives a brief introduction of feedback control systems. This consists of two main aspects. One aspect is the methodology of designing controllers for a nominal plant. The other is the description of model uncertainty that characterizes the difference between the actual plant and the nominal plant, and the analysis of stability of closed-loop systems with uncertainty.

As to the design of controllers for a nominal plant, first we have described a nominal system, and introduced a coprime factorization tool for the analysis of close-loop systems. Then for controller design we have given two fundamental conditions that are well-posedness and internal stability, ensuring the whole system is physically realizable and stable. Finally, we have discussed the YK parameterization of all stabilizing controllers for a given nominal plant, and have extended the results to the design of stabilizing 2-DOF controllers.

Two ways of representation of model uncertainty have been introduced, the coprime factor representation and dual YK parameterization. The former, combining with small gain theorem, provides an effective tool for the analysis of stability of closed-loop systems with uncertainty. The latter is, in comparison, more effective for controller synthesis.

3 Reinforcement Learning and Feedback Control

Reinforcement learning (RL) is a branch of machine learning that deals with making sequences of decisions. It refers to an agent that interacts with its environment, and receives an observation and reward. RL algorithms seek to maximize the agent's total reward, given an unknown environment, through a trial-and-error learning process [58, 65]. By contrast, feedback control is concerned with the design of a feedback controller for a given plant to optimize the system performance. The controller interacts with the plant by taking a control signal, and receives an observation and a cost. Oftentimes, an optimal controller can be found by minimizing the total cost. A simple interconnection of RL and feedback control is described in Fig. 3.1.

Note that as RL and feedback control are well developed by different communities, there is a critical terminology problem when it comes to describing optimization problems. This includes the use of agent/environment/reward/maximization in the former field and the use of controller/plant/cost/minimization in the latter field. A rather exhaustive list is given in [7] for the reader's reference.

From the concept of RL, one can perceive that RL has close connections to both optimal control and adaptive control. To be precise, RL refers to a class of model-free methods which enable the design of adaptive controllers that learn optimal solutions to user-prescribed optimal control problems [42]. To articulate the application of RL to feedback control, in this chapter, we will apply RL methods to solve two fundamental feedback control problems, the linear quadratic regulator (LQR) and the linear quadratic Gaussian (LQG).

We first give an overview of the state-of-the-art RL methods. Then we derive two dynamic programming based methods, policy iteration and value iteration, based on the Bellman equation and Bellman optimality equation, and apply them to solve the LQR and LQG problems. In addition, we elaborate two RL methods for solving the LQR problem. Finally, we develop a new RL method for solving the LQG problem, which is the key contribution of this chapter. This method is, to our best knowledge,

C. Hua, *Reinforcement Learning Aided Performance Optimization of Feedback Control Systems*, https://doi.org/10.1007/978-3-658-33034-7_3

the first try of efficiently solving a stochastic optimization problem in a partially observable environment.

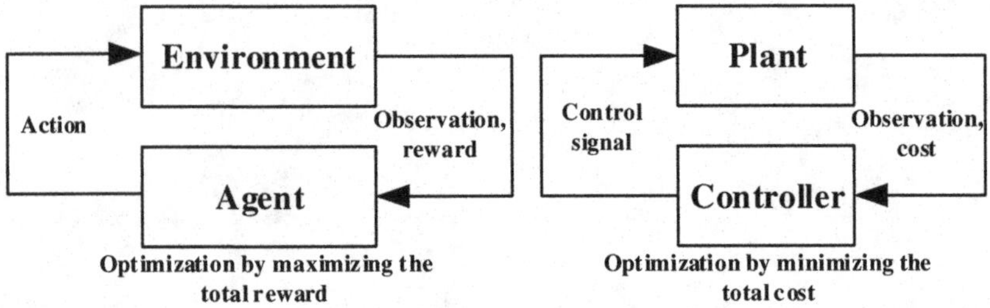

Figure 3.1 Interconnection of RL and feedback control

3.1 An Overview of RL Methods

Prior to the mathematical formulation of optimization problems, we give an intuitive view of the state-of-the-art RL methods. When RL is applied to controller optimization, there are two elements worth highlighting: 1) a controller (also called actor), whose control law (also called policy) is optimized during the learning phase; 2) a performance evaluator (also called critic) that evaluates the performance of the actor according to a prescribed cost function (also called value function). In accord to these two elements, RL methods can be classified into three categories [25]: actor-only, critic-only and actor-critic.

There are two ways to optimize a control policy using only an actor. First, there are derivative free optimization algorithms, including evolutionary algorithms. Their basic ideas are to perturb the policy parameters in different ways, such as in Gaussian distributions, and then to measure the performance and move the policy parameters to the direction in which the performance is optimized. These algorithms can work well for policies with a small amount of parameters, but scale poorly as the number of policy parameters increases. Some typical derivative free optimization algorithms are cross-entropy methods [67], covariance matrix adaptation [74] and natural evolution stategies [77]. Second, there are policy gradient algorithms [37, 54, 78]. Here we take a finite difference approach as an example. It lets policy parameters being varied with small increments and for each parameter variation an individual experiment is performed to measure the cost variation. Policy gradients are then estimated as the ratio of the cost variation to the parameter variation, and subsequently policy parameters are updated according to the policy gradients.

The second category of RL methods uses only a critic to optimize an actor. These RL algorithms focus on the learning of the critic, which not only evaluates how good the current actor is in terms of a prescribed cost function, but more importantly, it also reveals what a better actor can be. It should be emphasized that, although the actor is involved in the optimization, there is no direct parameterization of the actor and its optimization is fully guided by the critic. Some prevalent RL methods such as Q-learning [49] and state-action-reward-state-action (SARSA) [84] fall into this group. We shall also apply these RL methods to deal with the LQR problem and performance optimization problems of deterministic systems in the subsequent study.

The third approach for deriving RL methods uses both an actor and a critic. The control policy is parameterized by the actor. The critic is adopted to evaluate the performance of the actor. In addition, the critic provides policy gradients to the actor so that its policy parameters can be adjusted to improve system performance. Compared with the actor-only methods, actor-critic methods speed up learning significantly [58]. The natural actor-critic (NAC) method that deals with the LQG problem in this chapter, together with the state-of-the-art trust region policy optimization [59] and proximal policy optimization methods [60], are all examples of actor-critic methods.

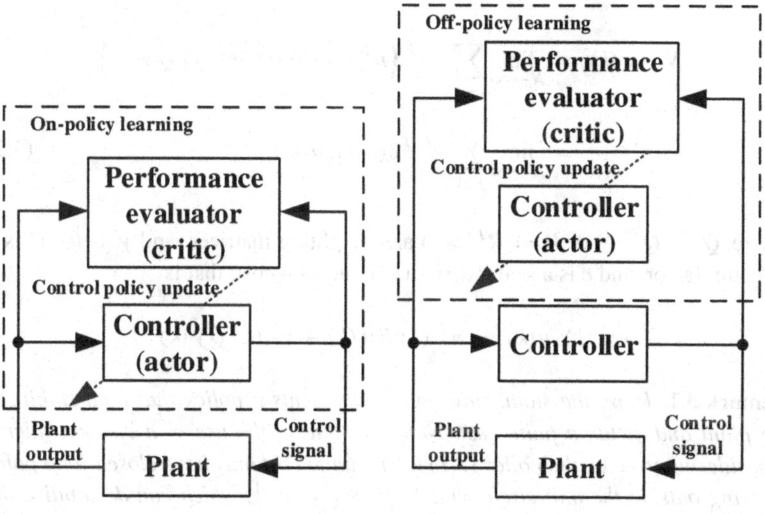

Figure 3.2 On-policy and off-policy RL for controller optimization

Finally, we classify RL methods alternatively as on-policy and off-policy algorithms, as we will extensively use these terms in the thesis. Their structures are shown in Fig. 3.2. Note that on-policy RL algorithms optimize an actor that regulates the plant, whereas off-policy RL algorithms optimize an actor, while another separate controller regulates the plant [65].

3.2 Infinite Horizon Linear Quadratic Regulator

In this section, we review the infinite horizon LQR problem from the viewpoint of both the dynamic programming (DP) and RL. It builds the foundation in our work on the application of RL methods to performance optimization of deterministic systems.

3.2.1 An Overview of the Infinite Horizon LQR Problem and DP

Consider the plant $\boldsymbol{P_0}$ (2.1), and a stabilizing and stationary policy

$$\boldsymbol{u}(k) = \mu\big(\boldsymbol{x}(k)\big) = \boldsymbol{F}\boldsymbol{x}(k) \tag{3.1}$$

with a state feedback gain $\boldsymbol{F}$, and define a performance index by a value (or cost) function over the infinite horizon

$$\begin{aligned} V^{\mu}\big(\boldsymbol{x}(k)\big) &= \lim_{N\to\infty} \sum_{i=k}^{N} \gamma^{i-k}\Big(\boldsymbol{u}^T(i)\,\boldsymbol{R}\boldsymbol{u}(i) + \boldsymbol{y}^T(i)\,\boldsymbol{Q}\boldsymbol{y}(i)\Big) \\ &= \lim_{N\to\infty} \sum_{i=k}^{N} \gamma^{i-k} c(\boldsymbol{x}(i), \boldsymbol{u}(i)), \end{aligned} \tag{3.2}$$

where $\boldsymbol{Q} = \boldsymbol{Q}^T \geq 0$, $\boldsymbol{R} = \boldsymbol{R}^T > 0$ are weighting matrices, and $\gamma \in (0, 1]$ is a discount factor, and c is a scalar denoting a one-step cost, that is

$$c(\boldsymbol{x}(k), \boldsymbol{u}(k)) = \boldsymbol{u}^T(k)\,\boldsymbol{R}\boldsymbol{u}(k) + \boldsymbol{y}^T(k)\,\boldsymbol{Q}\boldsymbol{y}(k)\,.$$

Remark 3.1 *Here, the stabilizing policy represents a policy that both stabilizes the plant and yields a finite cost. Note that due to the use of a discount factor (considering $\gamma < 1$), the policy (3.1) with a finite cost may have closed-loop poles z_p lying outside the unit circle with $1 \leq |z_p| < \gamma^{-1}$, which can destabilize the plant.*

The control objective is to find the optimal policy $\mu^*(\boldsymbol{x}(k)) = \boldsymbol{F}^*\boldsymbol{x}(k)$ with the optimal state feedback gain $\boldsymbol{F}^*$ that minimizes $V^{\mu}(\boldsymbol{x}(k))$. It can be described by

$$\mu^*(\boldsymbol{x}(k)) = \arg\min_{\boldsymbol{u}(k)} V^{\mu}(\boldsymbol{x}(k)). \tag{3.3}$$

This optimization problem is essentially an LQR problem. In order to find the optimal gain $\boldsymbol{F}^*$, we shall use the Bellman equation [5], a recursive form of (3.2), that is

$$V^{\mu}(\boldsymbol{x}(k)) = c(\boldsymbol{x}(k), \mu(\boldsymbol{x}(k))) + \gamma V^{\mu}(\boldsymbol{x}(k+1)), \tag{3.4}$$

and the Bellman optimality equation [5]

$$V^{\mu^*}(\boldsymbol{x}(k)) = \min_{\boldsymbol{u}(k)} \left(c(\boldsymbol{x}(k), \boldsymbol{u}(k)) + \gamma V^{\mu^*}(\boldsymbol{x}(k+1)) \right), \tag{3.5}$$

where $V^{\mu^*}(\boldsymbol{x}(k))$ is the value function of the optimal policy $\mu^*(\boldsymbol{x}(k))$. The above equation conveys a key property of an optimal policy that is it can be subdivided into two components: an optimal first control and a following optimal policy from the successor state. One can then use (3.5) to determine the optimal policy

$$\mu^*(\boldsymbol{x}(k)) = \arg\min_{\boldsymbol{u}(k)} \left(c(\boldsymbol{x}(k), \boldsymbol{u}(k)) + \gamma V^{\mu^*}(\boldsymbol{x}(k+1)) \right). \tag{3.6}$$

Note that Equation (3.6) provides a backward-in-time procedure for solving optimal control problems, since one must know the optimal control at $k+1$ in order to determine the optimal control at k. This is the key idea of DP to obtain the optimal policy.

In the LQR case, assume that for any stabilizing policy (3.1), the value function (3.3) can be alternatively represented as a quadratic form of the state, that is

$$V^{\mu}(\boldsymbol{x}(k)) = \boldsymbol{x}^T(k)\boldsymbol{P}^{\mu}\boldsymbol{x}(k), \tag{3.7}$$

with a symmetric kernel matrix $\boldsymbol{P}^{\mu} > 0$. Placing (3.7) back into (3.4), the Bellman equation for the LQR should be

$$\boldsymbol{x}^T(k)\boldsymbol{P}^{\mu}\boldsymbol{x}(k) = \boldsymbol{u}^T(k)\boldsymbol{R}\boldsymbol{u}(k) + \boldsymbol{y}^T(k)\boldsymbol{Q}\boldsymbol{y}(k) + \gamma\boldsymbol{x}^T(k+1)\boldsymbol{P}^{\mu}\boldsymbol{x}(k+1). \tag{3.8}$$

Substituting (2.1) and (3.1) into (3.8), and considering that it holds for all $\boldsymbol{x}(k)$, one has

$$\boldsymbol{P}^{\mu} = \boldsymbol{F}^T \boldsymbol{R} \boldsymbol{F} + (\boldsymbol{C} + \boldsymbol{D}\boldsymbol{F})^T \boldsymbol{Q} (\boldsymbol{C} + \boldsymbol{D}\boldsymbol{F}) + \gamma (\boldsymbol{A} + \boldsymbol{B}\boldsymbol{F})^T \boldsymbol{P}^{\mu} (\boldsymbol{A} + \boldsymbol{B}\boldsymbol{F}). \tag{3.9}$$

This is a Lyapunov equation, whose solution $\boldsymbol{P}^{\mu}$ justifies the form of the value function (3.7) for any stabilizing policy (3.1).

Here is how DP works for the LQR. Consider that the value function $V^{\mu^*}(\boldsymbol{x}(k))$ of the optimal policy $\mu^*(\boldsymbol{x}(k))$ has the form (3.7) with the kernel matrix $\boldsymbol{P}^{\mu^*}$. Based on (3.6), one can differentiate the terms in the outermost parentheses with respect to $\boldsymbol{u}(k)$ to obtain the optimal feedback gain

$$\boldsymbol{F}^* = -\left(\boldsymbol{R} + \boldsymbol{D}^T \boldsymbol{Q} \boldsymbol{D} + \gamma \boldsymbol{B}^T \boldsymbol{P}^{\mu^*} \boldsymbol{B}\right)^{-1} \left(\gamma \boldsymbol{B}^T \boldsymbol{P}^{\mu^*} \boldsymbol{A} + \boldsymbol{D}^T \boldsymbol{Q} \boldsymbol{C}\right). \tag{3.10}$$

Replacing $\boldsymbol{F}$ and $\boldsymbol{P}^{\mu}$ in (3.9) with $\boldsymbol{F}^*$ (3.10) and $\boldsymbol{P}^{\mu^*}$, Equation (3.9) becomes

$$\boldsymbol{P}^{\mu^*} = \left\{ \begin{array}{l} \gamma \boldsymbol{A}^T \boldsymbol{P}^{\mu^*} \boldsymbol{A} + \boldsymbol{C}^T \boldsymbol{Q} \boldsymbol{C} - \left(\gamma \boldsymbol{A}^T \boldsymbol{P}^{\mu^*} \boldsymbol{B} + \boldsymbol{C}^T \boldsymbol{Q} \boldsymbol{D}\right) \\ \times \left(\boldsymbol{R} + \boldsymbol{D}^T \boldsymbol{Q} \boldsymbol{D} + \gamma \boldsymbol{B}^T \boldsymbol{P}^{\mu^*} \boldsymbol{B}\right)^{-1} \left(\gamma \boldsymbol{B}^T \boldsymbol{P}^{\mu^*} \boldsymbol{A} + \boldsymbol{D}^T \boldsymbol{Q} \boldsymbol{C}\right) \end{array} \right\}. \tag{3.11}$$

This equation is known as the Riccati equation for the LQR. To obtain the optimal state feedback gain $\boldsymbol{F}^*$, we first solve the Riccati equation for $\boldsymbol{P}^{\mu^*}$, then place $\boldsymbol{P}^{\mu^*}$ into (3.10).

3.2.2 Policy Iteration and Value Iteration

In this subsection, we first elaborate two DP-based iterative methods: policy iteration and value iteration to obtain the optimal policy, then apply them to solving the infinite horizon LQR problem. Note that these two methods are of great importance and serve as the cornerstone of our further studies. Considering that Bellman equation (3.4) and Bellman optimality equations (3.5) are fixed point equations, we give the following policy evaluation and improvement lemma [8, 65].

Lemma 3.1 *Consider the plant* $\boldsymbol{P}_0$ *(2.1) and a stabilizing policy* $\mu(\boldsymbol{x}(k))$ *(3.1) with the value function* $V^{\mu}(\boldsymbol{x}(k))$ *(3.2). Start with any value* $V_0^{\mu}(\boldsymbol{x}(k))$ *for all* $\boldsymbol{x}$ *and iterate it, for* $j = 0, 1, \ldots,$ *with*

$$V_{j+1}^{\mu}(\boldsymbol{x}(k)) = c(\boldsymbol{x}(k), \mu(\boldsymbol{x}(k))) + \gamma V_j^{\mu}(\boldsymbol{x}(k+1)). \tag{3.12}$$

Then its limit has $\lim_{N\to\infty} V_N^{\mu}(\boldsymbol{x}(k)) = V^{\mu}(\boldsymbol{x}(k))$.

Lemma 3.2 *Given the plant* $\boldsymbol{P_0}$ *(2.1) and a stabilizing policy* $\mu(\boldsymbol{x}(k))$ *(3.1) with the value function* $V^{\mu}(\boldsymbol{x}(k))$ *(3.2), an improved policy* $\bar{\mu}\big(\boldsymbol{x}(k)\big)$ *is determined by*

$$\bar{\mu}(\boldsymbol{x}(k)) = \arg\min_{\boldsymbol{u}(k)} \big(c(\boldsymbol{x}(k), \boldsymbol{u}(k)) + V^{\mu}(\boldsymbol{x}(k+1))\big), \tag{3.13}$$

with the value function $V^{\bar{\mu}}(\boldsymbol{x}(k))$ *satisfying* $V^{\bar{\mu}}(\boldsymbol{x}(k)) \leq V^{\mu}(\boldsymbol{x}(k))$ *for all* $\boldsymbol{x}$*. And the equality sign holds if and only if both* $\mu(\boldsymbol{x}(k))$ *and* $\bar{\mu}(\boldsymbol{x}(k))$ *are optimal policies.*

It is evident that for any stabilizing policy $\mu(\boldsymbol{x}(k))$, its value function $V^{\mu}(\boldsymbol{x}(k))$ can be determined using policy evaluation by Lemma 3.1. Once $V^{\mu}(\boldsymbol{x}(k))$ is available, an improved policy can be derived using policy improvement by Lemma 3.2. Performing policy evaluation and policy improvement alternately until the optimal policy is reached gives the policy iteration algorithm. It is summarized in Algorithm 3.1.

Algorithm 3.1 Policy Iteration for Obtaining the Optimal Policy

---

1: **Initialization**: Select any stabilizing policy $\mu^0(\boldsymbol{x}(k))$ and initial value function $V_0^{\mu^0}(\boldsymbol{x}(k))$. Then for $i = 0, 1, \ldots$
2: **Repeat**
3: **Policy evaluation**: For $j = 0, 1, \ldots$
4: **Repeat**
5: $V_{j+1}^{\mu^i}(\boldsymbol{x}(k)) = c(\boldsymbol{x}(k), \boldsymbol{u}(k)) + \gamma V_j^{\mu^i}(\boldsymbol{x}(k+1))$
6: **Until** the value function converges, and set it as $V^{\mu^i}(\boldsymbol{x}(k))$
7: **Policy improvement**: Update the policy $\mu^i(\boldsymbol{x}(k))$ to $\mu^{i+1}(\boldsymbol{x}(k))$ with $\mu^{i+1}(\boldsymbol{x}(k)) = \operatorname{argmin}_{\boldsymbol{u}(k)} \Big(c(\boldsymbol{x}(k), \boldsymbol{u}(k)) + \gamma V^{\mu^i}(\boldsymbol{x}(k+1))\Big)$, and set $V_0^{\mu^{i+1}}(\boldsymbol{x}(k))$ as $V^{\mu^i}(\boldsymbol{x}(k))$.
8: **Until** the policy converges, and set it as the optimal policy $\mu^*(\boldsymbol{x}(k))$.

The downside of using policy iteration is that each of its iterations involves policy evaluation, which may itself be an elongated iterative computation. In fact, the policy evaluation step of policy iteration can be truncated without losing the convergence guarantees of policy iteration [65]. A typical case is when the policy evaluation stops only after applying (3.12) once at all states. This gives the value iteration algorithm described in Algorithm 3.2.

Algorithm 3.2 Value Iteration for Obtaining the Optimal Policy

1: **Initialization**: Select any stabilizing policy $\mu^0(\boldsymbol{x}(k))$ and initial value function $V^{\mu^0}(\boldsymbol{x}(k))$. Then for $i = 0, 1, \ldots$
2: **Repeat**
3: **Value update**: Update the value using
$V^{\mu^{i+1}}(\boldsymbol{x}(k)) = c(\boldsymbol{x}(k), \boldsymbol{u}(k)) + \gamma V^{\mu^i}(\boldsymbol{x}(k+1))$
4: **Policy improvement**: Update the policy $\mu^i(\boldsymbol{x}(k))$ to $\mu^{i+1}(\boldsymbol{x}(k))$ with
$\mu^{i+1}(\boldsymbol{x}(k)) = \underset{\boldsymbol{u}(k)}{\operatorname{argmin}} \left(c(\boldsymbol{x}(k), \boldsymbol{u}(k)) + \gamma V^{\mu^{i+1}}(\boldsymbol{x}(k+1))\right)$
5: **Until** the policy converges, and set it as the optimal policy $\mu^*(\boldsymbol{x}(k))$.

Remark 3.2 *Unlike policy iteration, there is no one-to-one mapping between intermediate value functions and policies in value iteration. The key idea of value iteration is to greedily update the existing value function using one-step look-ahead until it reaches the optimal one. Once the optimal value function is achieved, the greedy policy with respect to it delivers the optimal policy.*

Considering that in the LQR case, the value function of any stabilizing policy $\mu(\boldsymbol{x}(k))$ (3.1) has the form (3.7), the policy iteration and value iteration algorithms for estimating the optimal policy of the LQR problem are given in Algorithm 3.3 and Algorithm 3.4.

Algorithm 3.3 Policy Iteration for Estimating the Optimal Policy of the LQR

1: **Initialization**: Select any stabilizing policy $\mu^0(\boldsymbol{x}(k))$ (3.1) with $\boldsymbol{F}^0$ and initial value function $V_0^{\mu^0}(\boldsymbol{x}(k))$ (3.7) with a random kernel matrix $\boldsymbol{P}_0^{\mu^0}$. Then for $i = 0, 1, \ldots$
2: **Repeat**
3: **Policy evaluation**: for $j = 0, 1, \ldots$
4: **Repeat**
5: $\boldsymbol{P}_{j+1}^{\mu^i} = (\boldsymbol{F}^i)^T \boldsymbol{R} \boldsymbol{F}^i + (\boldsymbol{C}+\boldsymbol{D}\boldsymbol{F}^i)^T \boldsymbol{Q} (\boldsymbol{C}+\boldsymbol{D}\boldsymbol{F}^i) + \gamma (\boldsymbol{A}+\boldsymbol{B}\boldsymbol{F}^i)^T \boldsymbol{P}_j^{\mu^i} (\boldsymbol{A}+\boldsymbol{B}\boldsymbol{F}^i)$
6: **Until** the kernel matrix of the value function converges, and set it as $\boldsymbol{P}^{\mu^i}$.
7: **Policy improvement**: Update the policy $\mu^i(\boldsymbol{x}(k))$ to $\mu^{i+1}(\boldsymbol{x}(k))$ with $\boldsymbol{F}^{i+1} = -\left(\boldsymbol{R} + \boldsymbol{D}^T \boldsymbol{Q} \boldsymbol{D} + \gamma \boldsymbol{B}^T \boldsymbol{P}^{\mu^i} \boldsymbol{B}\right)^{-1} \left(\gamma \boldsymbol{B}^T \boldsymbol{P}^{\mu^i} \boldsymbol{A} + \boldsymbol{D}^T \boldsymbol{Q} \boldsymbol{C}\right)$, and set $\boldsymbol{P}_0^{\mu^{i+1}} = \boldsymbol{P}^{\mu^i}$.
8: **Until** the policy converges, and set it as the optimal policy $\mu^*(\boldsymbol{x}(k))$.

Note that DP-based methods assume a perfect model of the plant to compute the optimal policy. Although this assumption limits its utility, it gives insights on how the optimal policy can be achieved. In fact, RL methods introduced in the subsequent subsections only seek to achieve the same results as DP in a model-free manner.

Algorithm 3.4 Value Iteration for Estimating the Optimal Policy of the LQR

1: **Initialization**: Select any stabilizing policy $\mu^0(\boldsymbol{x}(k))$ (3.1) with $\boldsymbol{F}^0$ and initial value function $V^{\mu^0}(\boldsymbol{x}(k))$ (3.7) with a random kernel matrix $\boldsymbol{P}^{\mu^0}$. Then for $i = 0, 1, \ldots$
2: **Repeat**
3: **Value update**: Update the kernel matrix of the value function using
$\boldsymbol{P}^{\mu^{i+1}} = (\boldsymbol{F}^i)^T \boldsymbol{R} \boldsymbol{F}^i + (\boldsymbol{C}+\boldsymbol{D}\boldsymbol{F}^i)^T \boldsymbol{Q} (\boldsymbol{C}+\boldsymbol{D}\boldsymbol{F}^i) + \gamma (\boldsymbol{A}+\boldsymbol{B}\boldsymbol{F}^i)^T \boldsymbol{P}^{\mu^i} (\boldsymbol{A}+\boldsymbol{B}\boldsymbol{F}^i)$.
4: **Policy improvement**: Update the policy $\mu^i(\boldsymbol{x}(k))$ to $\mu^{i+1}(\boldsymbol{x}(k))$ with
$\boldsymbol{F}^{i+1} = -\left(\boldsymbol{R} + \boldsymbol{D}^T \boldsymbol{Q} \boldsymbol{D} + \gamma \boldsymbol{B}^T \boldsymbol{P}^{\mu^{i+1}} \boldsymbol{B}\right)^{-1} \left(\gamma \boldsymbol{B}^T \boldsymbol{P}^{\mu^{i+1}} \boldsymbol{A} + \boldsymbol{D}^T \boldsymbol{Q} \boldsymbol{C}\right)$.
5: **Until** the policy converges, and set it as the optimal policy $\mu^*(\boldsymbol{x}(k))$.

3.2.3 *Q*-learning

In this subsection, we shall show how the LQR problem can be solved using an off-policy critic-only RL method: Q-learning .

Q-learning, initiated by Watkins [76] in 1989, was one of the early breakthroughs in RL allowing an optimal policy to be learned in a model-free manner. It is devised based on a Q-function (also called action-value function) for a stabilizing policy $\mu(\boldsymbol{x}(k))$, defined as

$$Q^{\mu}(\boldsymbol{x}(k), \boldsymbol{u}(k)) = c(\boldsymbol{x}(k), \boldsymbol{u}(k)) + \gamma V^{\mu}(\boldsymbol{x}(k+1)). \tag{3.14}$$

Note that $Q^{\mu}(\boldsymbol{x}(k), \boldsymbol{u}(k))$ is the sum of the one-step cost incurred by taking a control $\boldsymbol{u}(k)$ at $\boldsymbol{x}(k)$, plus the total cost that would accure if the policy $\mu(\boldsymbol{x}(k))$ were followed from the state $\boldsymbol{x}(k+1)$. The control policy, that $\boldsymbol{u}(k)$ follows, is called a behavior policy denoted as $\mu_b(\boldsymbol{x}(k))$, and the policy $\mu(\boldsymbol{x}(k))$, that is to be learned, is called a target policy. The term "off-policy" means that $\mu_b(\boldsymbol{x}(k)) \neq \mu(\boldsymbol{x}(k))$ holds during the learning phase. Considering

$$Q^{\mu}(\boldsymbol{x}(k), \mu(\boldsymbol{x}(k))) = c(\boldsymbol{x}(k), \mu(\boldsymbol{x}(k))) + \gamma V^{\mu}(\boldsymbol{x}(k+1)) = V^{\mu}(\boldsymbol{x}(k)),$$

one can write $Q^{\mu}(\boldsymbol{x}(k), \boldsymbol{u}(k))$ in the following recursive form

$$Q^{\mu}\left(\boldsymbol{x}\left(k\right),\boldsymbol{u}\left(k\right)\right)=c(\boldsymbol{x}(k),\boldsymbol{u}(k))+\gamma Q^{\mu}\left(\boldsymbol{x}\left(k+1\right),\mu\left(\boldsymbol{x}(k+1)\right)\right). \quad (3.15)$$

It is evident that once $Q^{\mu}(\boldsymbol{x}(k),\boldsymbol{u}(k))$ can be identified using data, then an improved policy $\bar{\mu}\left(\boldsymbol{x}\left(k\right)\right)$ can be derived, according to Lemma 3.2, by

$$\bar{\mu}\left(\boldsymbol{x}\left(k\right)\right)=\arg\min_{\boldsymbol{u}(k)} Q^{\mu}(\boldsymbol{x}(k),\boldsymbol{u}(k)). \quad (3.16)$$

In the LQR case, the Q-function (3.15) can be explicitly represented, under the stabilizing policy $\mu\left(\boldsymbol{x}\left(k\right)\right)=\boldsymbol{F}\boldsymbol{x}(k)$, by

$$\begin{aligned} Q^{\mu}\left(\boldsymbol{x}\left(k\right),\boldsymbol{u}\left(k\right)\right)&=\boldsymbol{u}^{T}\left(k\right)\boldsymbol{R}\boldsymbol{u}\left(k\right)+\boldsymbol{y}^{T}\left(k\right)\boldsymbol{Q}\boldsymbol{y}\left(k\right)+\gamma\boldsymbol{x}^{T}\left(k+1\right)\boldsymbol{P}^{\mu}\boldsymbol{x}\left(k+1\right)\\ &=\begin{bmatrix}\boldsymbol{x}\left(k\right)\\ \boldsymbol{u}\left(k\right)\end{bmatrix}^{T}\begin{bmatrix}\gamma\boldsymbol{A}^{T}\boldsymbol{P}^{\mu}\boldsymbol{A}+\boldsymbol{C}^{T}\boldsymbol{Q}\boldsymbol{C} & \gamma\boldsymbol{A}^{T}\boldsymbol{P}^{\mu}\boldsymbol{B}+\boldsymbol{C}^{T}\boldsymbol{Q}\boldsymbol{D}\\ \gamma\boldsymbol{B}^{T}\boldsymbol{P}^{\mu}\boldsymbol{A}+\boldsymbol{D}^{T}\boldsymbol{Q}\boldsymbol{C} & \boldsymbol{R}+\boldsymbol{D}^{T}\boldsymbol{Q}\boldsymbol{D}+\gamma\boldsymbol{B}^{T}\boldsymbol{P}^{\mu}\boldsymbol{B}\end{bmatrix}\begin{bmatrix}\boldsymbol{x}\left(k\right)\\ \boldsymbol{u}\left(k\right)\end{bmatrix} \quad (3.17)\\ &=\begin{bmatrix}\boldsymbol{x}\left(k\right)\\ \boldsymbol{u}\left(k\right)\end{bmatrix}^{T}\begin{bmatrix}\boldsymbol{H}_{xx}^{\mu} & \boldsymbol{H}_{xu}^{\mu}\\ \boldsymbol{H}_{ux}^{\mu} & \boldsymbol{H}_{uu}^{\mu}\end{bmatrix}\begin{bmatrix}\boldsymbol{x}\left(k\right)\\ \boldsymbol{u}\left(k\right)\end{bmatrix}=\begin{bmatrix}\boldsymbol{x}\left(k\right)\\ \boldsymbol{u}\left(k\right)\end{bmatrix}^{T}\boldsymbol{H}^{\mu}\begin{bmatrix}\boldsymbol{x}\left(k\right)\\ \boldsymbol{u}\left(k\right)\end{bmatrix}\end{aligned}$$

where

$$\begin{aligned}\boldsymbol{H}^{\mu}&=\begin{bmatrix}\boldsymbol{H}_{xx}^{\mu} & \boldsymbol{H}_{xu}^{\mu}\\ \boldsymbol{H}_{ux}^{\mu} & \boldsymbol{H}_{uu}^{\mu}\end{bmatrix}\\ &=\begin{bmatrix}\gamma\boldsymbol{A}^{T}\boldsymbol{P}^{\mu}\boldsymbol{A}+\boldsymbol{C}^{T}\boldsymbol{Q}\boldsymbol{C} & \gamma\boldsymbol{A}^{T}\boldsymbol{P}^{\mu}\boldsymbol{B}+\boldsymbol{C}^{T}\boldsymbol{Q}\boldsymbol{D}\\ \gamma\boldsymbol{B}^{T}\boldsymbol{P}^{\mu}\boldsymbol{A}+\boldsymbol{D}^{T}\boldsymbol{Q}\boldsymbol{C} & \boldsymbol{R}+\boldsymbol{D}^{T}\boldsymbol{Q}\boldsymbol{D}+\gamma\boldsymbol{B}^{T}\boldsymbol{P}^{\mu}\boldsymbol{B}\end{bmatrix}\in\mathbb{R}^{(n+k_u)\times(n+k_u)}.\end{aligned}$$

Based on (3.16), the state feedback gain $\bar{\boldsymbol{F}}$ of the improved policy $\bar{\mu}$ is determined by

$$\bar{\boldsymbol{F}}=-(\boldsymbol{H}_{uu}^{\mu})^{-1}\boldsymbol{H}_{ux}^{\mu}=-\left(\boldsymbol{R}+\boldsymbol{D}^{T}\boldsymbol{Q}\boldsymbol{D}+\gamma\boldsymbol{B}^{T}\boldsymbol{P}^{\mu}\boldsymbol{B}\right)^{-1}\left(\gamma\boldsymbol{B}^{T}\boldsymbol{P}^{\mu}\boldsymbol{A}+\boldsymbol{D}^{T}\boldsymbol{Q}\boldsymbol{C}\right). \quad (3.18)$$

If the procedure of calculating the Q-function and improving the policy is done alternately, then the policy is guaranteed to converge to the optimal one for the LQR case [10]. We now articulate how the model-free Q-learning method approaches the optimal policy of the LQR problem using policy iteration. It has been shown in Algorithm 3.1 that policy iteration consists of two steps: policy evaluation and policy improvement. Note that policy improvement can be performed using (3.18) if $\boldsymbol{H}^{\mu}$ is available. Therefore the remaining issue is the evaluation of a given policy μ, whose core is the identification of $\boldsymbol{H}^{\mu}$.

Substitute (3.17) into (3.15), and re-arrange (3.15) in the following error form

$$\delta(k)=\begin{bmatrix}\boldsymbol{x}(k)\\ \boldsymbol{u}(k)\end{bmatrix}^T \boldsymbol{H}^{\mu}\begin{bmatrix}\boldsymbol{x}(k)\\ \boldsymbol{u}(k)\end{bmatrix}-\left(c(\boldsymbol{x}(k),\boldsymbol{u}(k))+\gamma\begin{bmatrix}\boldsymbol{x}(k+1)\\ \boldsymbol{F}\boldsymbol{x}(k+1)\end{bmatrix}^T \boldsymbol{H}^{\mu}\begin{bmatrix}\boldsymbol{x}(k+1)\\ \boldsymbol{F}\boldsymbol{x}(k+1)\end{bmatrix}\right) \tag{3.19}$$

$$=\boldsymbol{\psi}^T(k)\mathrm{vec}(\boldsymbol{H}^{\mu})-c(\boldsymbol{x}(k),\boldsymbol{u}(k)),$$

where $\delta(k)$ is called temporal difference (TD) error, and vec$(\cdot)$ is a vectorization operation that converts a matrix into a column vector and

$$\boldsymbol{\psi}(k)=\begin{bmatrix}\boldsymbol{x}(k)\\ \boldsymbol{u}(k)\end{bmatrix}\otimes\begin{bmatrix}\boldsymbol{x}(k)\\ \boldsymbol{u}(k)\end{bmatrix}-\gamma\begin{bmatrix}\boldsymbol{x}(k+1)\\ \boldsymbol{F}\boldsymbol{x}(k+1)\end{bmatrix}\otimes\begin{bmatrix}\boldsymbol{x}(k+1)\\ \boldsymbol{F}\boldsymbol{x}(k+1)\end{bmatrix},$$

where $\otimes$ denotes a Kronecker product operation. It can be seen from (3.19) that $\boldsymbol{H}^{\mu}$ can be identified by minimizing the TD error using available data $\boldsymbol{x}(k)$, $\boldsymbol{u}(k)$ and $\boldsymbol{x}(k+1)$. It is a linear regression problem. To solve it, stochastic gradient descent (SGD), batch or recursive least squares (LS) methods can be applied.

Considering an applied target policy $\mu^i(\boldsymbol{x}(k))$, we use a recursive LS method to identify the corresponding $\boldsymbol{H}^{\mu^i}$. Given the initial values $\boldsymbol{R}_0^i=\beta\boldsymbol{I}$ for some large constant β and $\boldsymbol{H}_0^{\mu^i}$, the recurrence relation is given by

$$\begin{cases}\mathrm{vec}(\boldsymbol{H}_{j+1}^{\mu^i})=\mathrm{vec}(\boldsymbol{H}_j^{\mu^i})+\dfrac{\boldsymbol{R}_j^i\boldsymbol{\psi}(k)\left(c(\boldsymbol{x}(k),\boldsymbol{u}(k))-\boldsymbol{\psi}^T(k)\,\mathrm{vec}(\boldsymbol{H}_j^{\mu^i})\right)}{1+\boldsymbol{\psi}^T(k)\boldsymbol{R}_j^i\boldsymbol{\psi}(k)}\\[2ex] \boldsymbol{R}_{j+1}^i=\boldsymbol{R}_j^i-\dfrac{\boldsymbol{R}_j^i\boldsymbol{\psi}(k)\boldsymbol{\psi}^T(k)\boldsymbol{R}_j^i}{1+\boldsymbol{\psi}^T(k)\boldsymbol{R}_j^i\boldsymbol{\psi}(k)},\end{cases} \tag{3.20}$$

where the subscript j denotes j^{th} estimate of $\boldsymbol{H}^{\mu^i}$, and the indices k and j are incremented at each time instant.

Goodwin [23] shows that the estimated parameters converge to the real ones if $\boldsymbol{H}^{\mu^i}$ is invariant and $\boldsymbol{\psi}(k)$ satisfies the following persistent excitation condition

$$\zeta_0\boldsymbol{I}\le\frac{1}{N}\sum_{l=1}^{N}\boldsymbol{\psi}(k-l)\boldsymbol{\psi}^T(k-l)\le\bar{\zeta}_0\boldsymbol{I},\quad\forall k\ge N_0 \text{ and } N\ge N_0,$$

where $\zeta_0 \leq \bar{\zeta}_0$, and N_0 is a positive number. The condition requires the behavior policy $\mu_b(\boldsymbol{x}(k))$ be different from the target policy. This attributes the off-policy property to the Q-learning. We can choose, in practice, the sum of the target policy plus white noise or simply white noise as the behavior policy $\mu_b(\boldsymbol{x}(k))$. Additionally, as $\boldsymbol{H}^{\mu^i}$ is a symmetric matrix having $(n+k_u)(n+k_u+1)/2$ unknown parameters, it can be identified with at least $(n+k_u)(n+k_u+1)/2$ data sets. Finally, we summarize the Q-learning policy iteration algorithm in Algorithm 3.5.

Algorithm 3.5 Q-learning: Policy Iteration for Estimating the Optimal Policy of the LQR

1: **Initialization**: Select any stabilizing policy $\mu^0(\boldsymbol{x}(k))$ (3.1) with $\boldsymbol{F}^0$ and initial Q-function $Q_0^{\mu^0}(\boldsymbol{x}(k))$ (3.17) with $\boldsymbol{H}_0^{\mu^0}$, and a large constant β. Then for $i = 0, 1, \ldots$
2: **Repeat**
3: Set $\boldsymbol{R}_0^i = \beta \boldsymbol{I}$.
4: **Policy evaluation**: for $j = 0, 1, \ldots$
5: **Repeat**
6: Update the estimation of $\boldsymbol{H}^{\mu^i}$ using (3.20) with data $\boldsymbol{x}, \boldsymbol{u}$.
7: **Until** the estimation converges, and set it as $\boldsymbol{H}^{\mu^i}$.
8: **Policy improvement**: Update the policy μ^i to μ^{i+1} with $\boldsymbol{F}^{i+1} = -(\boldsymbol{H}_{uu}^{\mu^i})^{-1}\boldsymbol{H}_{ux}^{\mu^i}$, and set $\boldsymbol{H}_0^{\mu^{i+1}} = \boldsymbol{H}^{\mu^i}$.
9: **Until** the policy converges, and set it as the optimal policy μ^*.

3.2.4 SARSA

In this subsection, we shall show how an on-policy RL method, SARSA, learns the optimal policy of the LQR based on a value function.

To explain the principle of SARSA, we choose for the nominal plant $\boldsymbol{P}_0$ (2.19), different from (3.1), a Gaussian stochastic and stabilizing policy

$$\boldsymbol{u}(k) = \pi(\boldsymbol{x}(k)) = \boldsymbol{F}\boldsymbol{x}(k) + \boldsymbol{\xi}(k) \tag{3.21}$$

with a state feedback gain $\boldsymbol{F}$ and Gaussian white noise $\boldsymbol{\xi} \sim N(\boldsymbol{0}, \boldsymbol{\Sigma}_{\xi})$. Associated with the policy, a value function is defined by

$$
\begin{aligned}
V^{\pi}\left(\boldsymbol{x}\left(k\right)\right) &= \lim_{N\to\infty} \mathbb{E}\left(\sum_{i=k}^{N} \gamma^{i-k}\left(\boldsymbol{u}^{T}\left(i\right)\boldsymbol{R}\boldsymbol{u}\left(i\right)+\boldsymbol{y}^{T}\left(i\right)\boldsymbol{Q}\boldsymbol{y}\left(i\right)\right)\right) \\
&= \lim_{N\to\infty} \mathbb{E}\left(\sum_{i=k}^{N} \gamma^{i-k} c(\boldsymbol{x}(i),\boldsymbol{u}(i))\right),
\end{aligned}
\tag{3.22}
$$

where the discount factor γ has $\gamma \in (0, 1)$, and $\mathbb{E}(\cdot)$ represents the expectation taken over all possible control inputs that are sampled from the stochastic policy π from the time instant k onward.

Remark 3.3 *Due to the use of the stochastic policy π, the discount factor γ in the value function (3.22) should not equal 1, otherwise all stabilizing policies will have infinite values and learning cannot be performed.*

The Bellman equation can be represented by

$$
V^{\pi}\left(\boldsymbol{x}\left(k\right)\right) = \mathbb{E}_{\boldsymbol{u}(k)}\left(c(\boldsymbol{x}(k),\boldsymbol{u}\left(k\right)) + \gamma V^{\pi}\left(\boldsymbol{x}\left(k+1\right)\right)\right), \tag{3.23}
$$

where $\mathbb{E}_{\boldsymbol{u}(k)}(\cdot)$ represents the expectation taken over $\boldsymbol{u}(k)$. Assume that for any policy (3.21), the value function (3.22) can be equivalently represented by

$$
V^{\pi}\left(\boldsymbol{x}\left(k\right)\right) = \boldsymbol{x}^{T}\left(k\right)\boldsymbol{P}^{\pi}\boldsymbol{x}\left(k\right) + c^{\pi}, \tag{3.24}
$$

with a symmetric kernel matrix $\boldsymbol{P}^{\pi} > 0$ and a constant c^{π}. Substituting (3.24) into (3.23) and considering the dynamics of the plant $\boldsymbol{P}_0$ (2.19), it can be derived that $\boldsymbol{P}^{\pi}$ is the solution to the Lyapunov equation (3.9), where $\boldsymbol{P}^{\pi}$ equals $\boldsymbol{P}^{\mu}$, and c^{π} is a constant that is

$$
c^{\pi} = \frac{1}{1-\gamma}\mathrm{tr}\left(\left(\gamma\boldsymbol{B}^{T}\boldsymbol{P}^{\pi}\boldsymbol{B} + \boldsymbol{R} + \boldsymbol{D}^{T}\boldsymbol{Q}\boldsymbol{D}\right)\boldsymbol{\Sigma}_{\xi}\right), \tag{3.25}
$$

where $\mathrm{tr}(\cdot)$ denotes a trace operation. This justifies the form of the value function (3.24) for any stabilizing policy (3.21). According to the DP introduced in Subsection 3.2.1, one can readily derive that the optimal policy π^{*} has the same optimal feedback gain $\boldsymbol{F}^{*}$ as the one in the noise-free case given by (3.10). This is a key result, as we will use the stochastic policy π to learn the optimal policy in SARSA.

SARSA resembles Q-learning, which uses also Q-function. It is defined for a stabilizing policy $\pi^{i}\left(\boldsymbol{x}\left(k\right)\right)$ by

$$Q^{\pi^i}(\boldsymbol{x}(k), \boldsymbol{u}(k)) = c(\boldsymbol{x}(k), \boldsymbol{u}(k)) + \gamma V^{\pi^i}(\boldsymbol{x}(k+1)). \tag{3.26}$$

The key difference between SARSA and Q-learning is that in SARSA, $\boldsymbol{u}(k)$ follows the target policy $\pi^i(\boldsymbol{x}(k))$. In the LQR case, similar to (3.17), one can represent the Q-function by

$$\begin{aligned} Q^{\pi^i}(\boldsymbol{x}(k), \boldsymbol{u}(k)) &= \begin{bmatrix} \boldsymbol{x}(k) \\ \boldsymbol{u}(k) \end{bmatrix}^T \begin{bmatrix} \boldsymbol{H}^{\pi^i}_{xx} & \boldsymbol{H}^{\pi^i}_{xu} \\ \boldsymbol{H}^{\pi^i}_{ux} & \boldsymbol{H}^{\pi^i}_{uu} \end{bmatrix} \begin{bmatrix} \boldsymbol{x}(k) \\ \boldsymbol{u}(k) \end{bmatrix} + c_h^{\pi^i} \\ &= \begin{bmatrix} \boldsymbol{x}(k) \\ \boldsymbol{u}(k) \end{bmatrix}^T \boldsymbol{H}^{\pi^i} \begin{bmatrix} \boldsymbol{x}(k) \\ \boldsymbol{u}(k) \end{bmatrix} + c_h^{\pi^i}, \end{aligned} \tag{3.27}$$

where $\boldsymbol{H}^{\pi^i}$ is a constant matrix that relates to $\boldsymbol{P}^{\pi^i}$ and $c_h^{\pi^i}$ is a constant that relates to c^{π^i}. The optimal policy can then be achieved using the policy iteration as adopted by Q-learning. Recursive LS can be applied to identify $\boldsymbol{H}^{\pi^i}$ and $c_h^{\pi^i}$ by minimizing the TD error [65]

$$\begin{aligned} \delta(k) =& Q^{\pi^i}(\boldsymbol{x}(k), \boldsymbol{u}(k)) - \left(c(\boldsymbol{x}(k), \boldsymbol{u}(k)) + Q^{\pi^i}(\boldsymbol{x}(k+1), \boldsymbol{u}(k+1))\right) \\ =& \begin{bmatrix} \boldsymbol{x}(k) \\ \boldsymbol{u}(k) \end{bmatrix}^T \boldsymbol{H}^{\pi^i} \begin{bmatrix} \boldsymbol{x}(k) \\ \boldsymbol{u}(k) \end{bmatrix} - \gamma \begin{bmatrix} \boldsymbol{x}(k+1) \\ \boldsymbol{u}(k+1) \end{bmatrix}^T \boldsymbol{H}^{\pi^i} \begin{bmatrix} \boldsymbol{x}(k+1) \\ \boldsymbol{u}(k+1) \end{bmatrix} \\ &+ (1-\gamma) c_h^{\pi^i} - c(\boldsymbol{x}(k), \boldsymbol{u}(k)). \end{aligned} \tag{3.28}$$

Note that here in (3.28) both $\boldsymbol{u}(k)$ and $\boldsymbol{u}(k+1)$ are sampled from the same policy π^i. Once $\boldsymbol{H}^{\pi^i}$ is available, a better policy with the feedback gain that equals $-(\boldsymbol{H}^{\pi^i}_{uu})^{-1}\boldsymbol{H}^{\pi^i}_{ux}$ can be achieved. Running this procedure repeatedly until the policy converges delivers the optimal policy.

However, a critical problem with SARSA is that the identified $\boldsymbol{H}^{\pi^i}$ using (3.28) has a high variance because of the use of the stochastic policy. This can easily lead to a wrong policy update, and subsequently system instability. To lower the variance, we suggest to use, instead of (3.28), the following equation to iteratively identify $\boldsymbol{H}^{\pi^i}$

$$\begin{aligned} \delta_j(k) =& Q^{\pi^i}_{j+1}(\boldsymbol{x}(k), \boldsymbol{u}(k)) - (1-\lambda) Q^{\pi^i}_j(\boldsymbol{x}(k), \boldsymbol{u}(k)) \\ &- \lambda \left(c(\boldsymbol{x}(k), \boldsymbol{u}(k)) + \gamma Q^{\pi^i}_j(\boldsymbol{x}(k+1), \boldsymbol{u}(k+1))\right) \end{aligned} \tag{3.29}$$

with a randomly chosen $Q_0^{\pi^i}$, where $\lambda \in (0, 1]$ is a learning rate. A smaller λ is preferred for a smoother policy update but at cost of slow learning. Finally, the SARSA policy iteration algorithm is summarized in Algorithm 3.6.

Algorithm 3.6 SARSA: Policy Iteration for the Optimal Policy of the LQR

1: **Initialization**: Select any stabilizing policy $\pi^0(\boldsymbol{x}(k))$ (3.1) with $\boldsymbol{F}^0$ and an initial Q-function $Q_0^{\pi^0}(\boldsymbol{x}(k))$ (3.27) with a random kernel matrix $\boldsymbol{H}_0^{\pi^0}$ and a constant $c_{h.0}^{\pi^0}$. Then for $i = 0, 1, \ldots$

2: **Repeat**

3: **Policy evaluation**: for $j = 0, 1, \ldots$

4: **Repeat**

5: Identify $\boldsymbol{H}_{j+1}^{\pi^i}$ and $c_{h.j+1}^{\pi^i}$ using LS or SGD by minimizing δ_j (3.28) with data $\boldsymbol{x}$, $\boldsymbol{u}$, and $\boldsymbol{H}_j^{\pi^i}$ and $c_{h.j}^{\pi^i}$.

6: **Until** $\boldsymbol{H}_j^{\pi^i}$ converges, and set it as $\boldsymbol{H}^{\pi^i}$, and set the constant as $c_h^{\pi^i}$.

7: **Policy improvement**: Update the policy π^i to π^{i+1} with $\boldsymbol{F}^{i+1} = -(\boldsymbol{H}_{\boldsymbol{uu}}^{\pi^i})^{-1}\boldsymbol{H}_{\boldsymbol{ux}}^{\pi^i}$, and set $\boldsymbol{H}_0^{\pi^{i+1}} = \boldsymbol{H}^{\pi^i}$ and $c_{h.0}^{\pi^{i+1}} = c_h^{\pi^i}$.

8: **Until** the policy converges, and set it as the optimal policy π^*.

In conclusion, both Q-learning and SARSA are critic-only RL methods. They both revolve around a Q-function for policy improvement and are guaranteed to achieve the optimal policy for the LQR problem. The fundamental difference is that Q-learning is an off-policy method and SARSA is an on-policy method.

3.2.5 Simulation Results

In this subsection, Q-learning and SARSA methods are applied to learn an LQR of a single-input single-output (SISO) LTI plant P_0, and their results are compared. The plant is described in the following continuous-time state-space form

$$\dot{\boldsymbol{x}} = \boldsymbol{A}\boldsymbol{x} + \boldsymbol{b}u, \quad y = \boldsymbol{c}\boldsymbol{x},$$

$$\boldsymbol{A} = \begin{bmatrix} 0.9376 & -0.0097 \\ 0.0774 & 0.9976 \end{bmatrix}, \boldsymbol{b} = \begin{bmatrix} 0.0242 \\ 0.00098 \end{bmatrix}, \boldsymbol{c} = \begin{bmatrix} 0 & 1 \end{bmatrix}.$$

The sampling time is set to 1 ms. The cost function in the Q-learning case is defined in the form of (3.2) and in the SARSA case in the form of (3.22) with the following weighting factors $R = 0.1$, $Q = 1$, and with the discount factor $\gamma = 0.99$. Theoretically, the optimal state feedback gain is calculated as $\boldsymbol{F}^* = \begin{bmatrix}-1.9932 & -2.3266\end{bmatrix}$. Fig. 3.3 shows the comparison of state feedback gains of the learned LQR controllers by Q-learning and SARSA, respectively. It is clear that Q-learning shows superior performance over SARSA in terms of accuracy and speed of convergence.

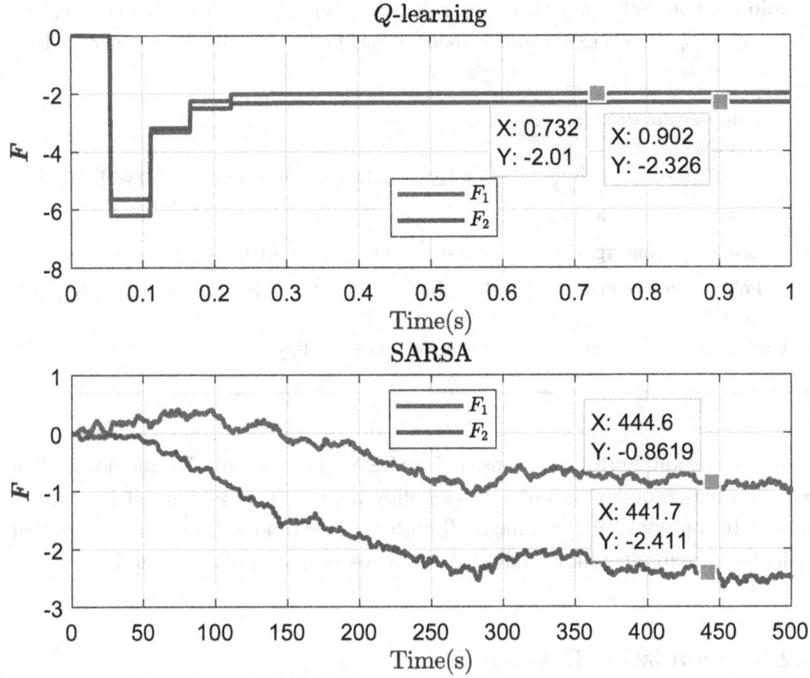

Figure 3.3 Comparison of Q-learning and SARSA

3.3 Infinite Horizon Linear Quadratic Gaussian

In this section, we review a linear quadratic Gaussian (LQG) control problem and provide solutions to it from the viewpoint of both stochastic DP and RL. It builds the foundation in our work on the application of RL methods to performance optimization of stochastic systems.

3.3.1 An Overview of Infinite Horizon LQG Problem and Stochastic DP

We consider a stochastic plant $\boldsymbol{P}$ that is extended from the nominal plant $\boldsymbol{P_0}$ (2.1), described by

$$\boldsymbol{P}: \begin{cases} \boldsymbol{x}(k+1) = \boldsymbol{A}\boldsymbol{x}(k) + \boldsymbol{B}\big(\boldsymbol{u}(k) + \boldsymbol{\omega}(k)\big), \quad \boldsymbol{x}(0) = \boldsymbol{x}_0 \\ \boldsymbol{y}(k) = \boldsymbol{C}\boldsymbol{x}(k) + \boldsymbol{D}\big(\boldsymbol{u}(k) + \boldsymbol{\omega}(k)\big) + \boldsymbol{v}(k), \end{cases} \tag{3.30}$$

where $\boldsymbol{\omega}(k)$, $\boldsymbol{v}(k)$ are actuator and measurement Gaussian white noises having $\boldsymbol{\omega}(k) \sim N(\boldsymbol{0}, \boldsymbol{\Sigma}_\omega)$ and $\boldsymbol{v}(k) \sim N(\boldsymbol{0}, \boldsymbol{\Sigma}_v)$. They are statistically independent of $\boldsymbol{u}(k)$ and $\boldsymbol{x}_0$.

Considering that a stabilizing nominal observer-based state feedback controller $\boldsymbol{K_0}$ (2.4) is designed for $\boldsymbol{P}$, the error dynamics is governed by

$$\begin{cases} \tilde{\boldsymbol{x}}(k+1) = \boldsymbol{A_L}\tilde{\boldsymbol{x}}(k) + \boldsymbol{B_L}\boldsymbol{\omega}(k) - \boldsymbol{L}\boldsymbol{v}(k), \\ \boldsymbol{r}(k) = \boldsymbol{C}\tilde{\boldsymbol{x}}(k) + \boldsymbol{D}\boldsymbol{\omega}(k) + \boldsymbol{v}(k), \end{cases} \tag{3.31}$$

where $\tilde{\boldsymbol{x}}(k) = \boldsymbol{x}(k) - \hat{\boldsymbol{x}}(k)$ is the state estimation error, and $\boldsymbol{A_L} = \boldsymbol{A} - \boldsymbol{LC}$ and $\boldsymbol{B_L} = \boldsymbol{B} - \boldsymbol{LD}$. Here we denote the control policy of $\boldsymbol{K_0}$ as $\mu(\hat{\boldsymbol{x}}(k))$ that is

$$\boldsymbol{u}(k) = \mu(\hat{\boldsymbol{x}}(k)) = \boldsymbol{F}\hat{\boldsymbol{x}}(k), \tag{3.32}$$

and associated with it, a performance index is defined by the following value function

$$\begin{aligned} V^\mu\big(\boldsymbol{x}(k)\big) &= \lim_{N\to\infty} \mathbb{E}\left(\sum_{i=k}^{N} \gamma^{i-k}\left(\boldsymbol{u}^T(i)\,\boldsymbol{R}\boldsymbol{u}(i) + \boldsymbol{y}^T(i)\,\boldsymbol{Q}\boldsymbol{y}(i)\right)\right) \\ &= \lim_{N\to\infty} \mathbb{E}\left(\sum_{i=k}^{N} \gamma^{i-k} c(\boldsymbol{x}(i), \boldsymbol{u}(i))\right) \end{aligned} \tag{3.33}$$

with $\gamma \in (0, 1)$. The infinite horizon LQG problem is then formulated as: given the plant model $\boldsymbol{P}$ (3.30) and the observer-based controller $\boldsymbol{K_0}$ (2.4), find the optimal policy μ^* with the feedback gain $\boldsymbol{F}^*$ and the observer gain $\boldsymbol{L}^*$ such that the value function is minimized.

To achieve the optimal policy, stochastic DP is used. We shall show the optimal value at time instant k has the quadratic form [61, 71]

$$V^{\mu^*}(\boldsymbol{x}(k)) = \boldsymbol{x}^T(k)\,\boldsymbol{P}^{\mu^*}(k)\boldsymbol{x}(k) + \tilde{\boldsymbol{x}}^T(k)\,\boldsymbol{W}^{\mu^*}(k)\tilde{\boldsymbol{x}}(k) + c^{\mu^*}(k), \tag{3.34}$$

where $\boldsymbol{P}^{\mu^*}(k)$, $\boldsymbol{W}^{\mu^*}(k)$ and $c^{\mu^*}(k)$ are parameters of the optimal value. Considering the final time instant $N(N \to \infty)$, clearly the optimal control $\boldsymbol{u}(N)$ is $\boldsymbol{0}$. Then the optimal value at N is

$$V^{\mu^*}(\boldsymbol{x}(N)) = \boldsymbol{x}^T(N)\,\boldsymbol{P}^{\mu^*}(N)\,\boldsymbol{x}(N) + c^{\mu^*}(N) \tag{3.35}$$

with $\boldsymbol{P}^{\mu^*}(N) = \boldsymbol{C}^T\boldsymbol{Q}\boldsymbol{C}$, $\boldsymbol{W}^{\mu^*} = \boldsymbol{0}$ and $c^{\mu^*}(N) = \mathrm{tr}(\boldsymbol{\Sigma}_\omega \boldsymbol{D}^T\boldsymbol{Q}\boldsymbol{D}) + \mathrm{tr}(\boldsymbol{\Sigma}_v\boldsymbol{Q})$. Now we suppose at $k+1$, the optimal value has the form (3.34) and prove the optimal value at k has the same form. Considering the dynamics of $\boldsymbol{x}(k+1)$ (3.30) and $\tilde{\boldsymbol{x}}(k+1)$ (3.31), the Q-function at k, when a control $\boldsymbol{u}(k)$ is taken, can be described by

$$\begin{aligned} &Q(\boldsymbol{x}(k), \boldsymbol{u}(k)) = \mathbb{E}\left(c(\boldsymbol{x}(k), \boldsymbol{u}(k)) + \gamma V^{\mu^*}(\boldsymbol{x}(k+1))\right) \\ &= \left\{ \begin{array}{l} \begin{bmatrix} \boldsymbol{x}(k) \\ \boldsymbol{u}(k) \end{bmatrix}^T \begin{bmatrix} \gamma \boldsymbol{A}^T\boldsymbol{P}^{\mu^*}(k+1)\boldsymbol{A} + \boldsymbol{C}^T\boldsymbol{Q}\boldsymbol{C} & \gamma\boldsymbol{A}^T\boldsymbol{P}^{\mu^*}(k+1)\boldsymbol{B} + \boldsymbol{C}^T\boldsymbol{Q}\boldsymbol{D} \\ \gamma\boldsymbol{B}^T\boldsymbol{P}^{\mu^*}(k+1)\boldsymbol{A} + \boldsymbol{D}^T\boldsymbol{Q}\boldsymbol{C} & \boldsymbol{R} + \boldsymbol{D}^T\boldsymbol{Q}\boldsymbol{D} + \gamma\boldsymbol{B}^T\boldsymbol{P}^{\mu^*}(k+1)\boldsymbol{B} \end{bmatrix} \begin{bmatrix} \boldsymbol{x}(k) \\ \boldsymbol{u}(k) \end{bmatrix} \\ +\gamma\tilde{\boldsymbol{x}}^T(k)\,\boldsymbol{A}_L^T\boldsymbol{W}^{\mu^*}(k+1)\,\boldsymbol{A}_L\tilde{\boldsymbol{x}}(k) + \mathrm{tr}\left(\boldsymbol{\Sigma}_\omega\boldsymbol{D}^T\boldsymbol{Q}\boldsymbol{D}\right) + \mathrm{tr}\left(\boldsymbol{\Sigma}_v\boldsymbol{Q}\right) \\ +\gamma\left(\mathrm{tr}\left(\boldsymbol{P}^{\mu^*}(k+1)\,\boldsymbol{B}\boldsymbol{\Sigma}_\omega\boldsymbol{B}^T + \boldsymbol{W}^{\mu^*}(k+1)\left(\boldsymbol{B}_L\boldsymbol{\Sigma}_\omega\boldsymbol{B}_L^T + \boldsymbol{L}\boldsymbol{\Sigma}_v\boldsymbol{L}^T\right)\right)\right) + \gamma c^{\mu^*}(k+1) \end{array} \right\}. \end{aligned} \tag{3.36}$$

Taking the derivative of the above Q-function with respect to $\boldsymbol{u}(k)$, the optimal control at k can be attained. However, it is a function of $\boldsymbol{x}(k)$ that is inaccessible to us. We shall use its unbiased estimation $\hat{\boldsymbol{x}}(k)$ instead. Finally, the optimal control at k is

$$\begin{aligned} \boldsymbol{u}(k) = \mu^*(\hat{\boldsymbol{x}}(k)) = &-\left(\boldsymbol{R} + \boldsymbol{D}^T\boldsymbol{Q}\boldsymbol{D} + \gamma\boldsymbol{B}^T\boldsymbol{P}^{\mu^*}(k+1)\boldsymbol{B}\right)^{-1} \\ &\times\left(\gamma\boldsymbol{B}^T\boldsymbol{P}^{\mu^*}(k+1)\boldsymbol{A} + \boldsymbol{D}^T\boldsymbol{Q}\boldsymbol{C}\right)\hat{\boldsymbol{x}}(k). \end{aligned} \tag{3.37}$$

Considering that

$$\begin{bmatrix} \boldsymbol{x}(k) \\ \hat{\boldsymbol{x}}(k) \end{bmatrix}^T \begin{bmatrix} \boldsymbol{0} & -\boldsymbol{Z} \\ -\boldsymbol{Z} & \boldsymbol{Z} \end{bmatrix} \begin{bmatrix} \boldsymbol{x}(k) \\ \hat{\boldsymbol{x}}(k) \end{bmatrix} = \begin{bmatrix} \boldsymbol{x}(k) \\ \tilde{\boldsymbol{x}}(k) \end{bmatrix}^T \begin{bmatrix} -\boldsymbol{Z} & \boldsymbol{0} \\ \boldsymbol{0} & \boldsymbol{Z} \end{bmatrix} \begin{bmatrix} \boldsymbol{x}(k) \\ \tilde{\boldsymbol{x}}(k) \end{bmatrix}$$

holds for any symmetric matrix $\boldsymbol{Z}$ and substituting (3.37) into (3.36), the optimal value at k can be derived, which has the form of (3.34) with

$$P^{\mu^*}(k)=\left\{\begin{array}{l}\gamma A^T P^{\mu^*}(k+1)A+C^T QC-\left(\gamma A^T P^{\mu^*}(k+1)B+C^T QD\right)\\ \times\left(R+D^T QD+\gamma B^T P^{\mu^*}(k+1)B\right)^{-1}\left(\gamma B^T P^{\mu^*}(k+1)A+D^T QC\right)\end{array}\right\}$$

$$W^{\mu^*}(k)=\left\{\begin{array}{l}\gamma A_L^T W^{\mu^*}(k+1)\,A_L+\left(\gamma A^T P^{\mu^*}(k+1)B+C^T QD\right)\\ \times\left(R+D^T QD+\gamma B^T P^{\mu^*}(k+1)B\right)^{-1}\left(\gamma B^T P^{\mu^*}(k+1)A+D^T QC\right)\end{array}\right\}$$

$$c^{\mu^*}(k)=\left\{\begin{array}{l}\gamma\left(\mathrm{tr}\left(P^{\mu^*}(k+1)\,B\Sigma_\omega B^T+W^{\mu^*}(k+1)\left(B_L\Sigma_\omega B_L^T+L\Sigma_v L^T\right)\right)\right)\\ +\mathrm{tr}\left(\Sigma_\omega D^T QD\right)+\mathrm{tr}\,(\Sigma_v Q)+\gamma c^{\mu^*}(k+1)\end{array}\right\}. \tag{3.38}$$

This justifies the assumed form of the optimal value function (3.34). It is evident that if the optimal control is computed recursively backwards-in-time, the state feedback gain converges to the optimal state feedback gain F^* as given in (3.10) and meanwhile the converged P^{μ^*} is the solution to the Riccati equation as given in (3.11).

Remark 3.4 *For any stabilizing observer gain* L *of* K_0 *(not necessary optimal), the value function has the form (3.34) as long as the state feedback gain is optimal.*

To determine the optimal observer gain at k, we first compute the expectation of the value function (3.34) with parameters in (3.38) and select out all terms including L and assemble them as a function of L expressed by

$$h(L)=\gamma\mathrm{tr}\left(W^{\mu^*}(k+1)\left(A_L\Sigma_{\tilde{x}}(k)A_L^T+B_L\Sigma_\omega B_L^T+L\Sigma_v L^T\right)\right), \tag{3.39}$$

where $\Sigma_{\tilde{x}}(k)=E(\tilde{x}(k)\tilde{x}^T(k))$ denotes the variance of $\tilde{x}$ at k. Then the derivative of $h(L)$ with respect to L is computed and set to $\mathbf{0}$, the optimal observer gain at k is determined, irrespective of $W^{\mu^*}(k+1)$, as

$$L^*(k)=\left(A\Sigma_{\tilde{x}}(k)\,C^T+B\Sigma_\omega D^T\right)\left(C\Sigma_{\tilde{x}}(k)\,C^T+D\Sigma_\omega D^T+\Sigma_v\right)^{-1}. \tag{3.40}$$

Clearly, $L^*(k)$ depends only on $\Sigma_{\tilde{x}}(k)$ that can be computed, based on (3.31), by

$$\begin{gathered}\Sigma_{\tilde{x}}(k)=\left\{\begin{array}{l}A_{L^*}(k-1)\Sigma_{\tilde{x}}(k-1)\,A_{L^*}^T(k-1)+B_{L^*}(k-1)\Sigma_\omega B_{L^*}^T(k-1)\\ +L^*(k-1)\Sigma_v L^*(k-1)^T\end{array}\right\},\\ A_{L^*}(k-1)=A-L^*(k-1)C,\ B_{L^*}(k-1)=B-L^*(k-1)D.\end{gathered} \tag{3.41}$$

In the context of infinite horizon, if (3.40) and (3.41) are computed recursively forwards-in-time, $\boldsymbol{\Sigma}_{\tilde{x}}(k)$ converges to a constant matrix and thus $\boldsymbol{L}^*(k)$ converges, which gives the optimal gain $\boldsymbol{L}^*$, i.e., the Kalman filter gain.

As is known, stochastic DP is a model-based optimization tool. Although in practice, a lack of a perfect plant model restricts its application, it provides insights on optimal solutions in the stochastic setting. In what follows, we will attempt to achieve an optimal solution using RL.

3.3.2 On-policy Natural Actor-critic

To deal with the LQG problem, we consider, in this subsection, on-policy policy gradient methods. Their intuitive idea is that, first a control policy of a feedback controller is parameterized with some parameters, then the policy is optimized based on the gradient of the value function with respect to the policy parameters.

Interested readers are referred to [54, 59, 63, 65] for overview of various policy gradient methods. Among them, the natural actor-critic (NAC) is the focus of the study. We shall show how it can be used to approach the LQG problem.

(A). Parameterization of the Control Policy

We consider the plant $\boldsymbol{P}$ (3.30) and parameterize a general output feedback controller $\boldsymbol{K} \in \mathcal{R}_{sp}$ by

$$\boldsymbol{u}(k) = \begin{bmatrix} \boldsymbol{\Theta}_{u.l} \\ \boldsymbol{\Theta}_{y.l} \end{bmatrix}^T \begin{bmatrix} \boldsymbol{u}_l(k-1) \\ \boldsymbol{y}_l(k-1) \end{bmatrix} + \boldsymbol{\xi}(k) = \boldsymbol{\Theta}^T \boldsymbol{z}_{uy.l}(k-1) + \boldsymbol{\xi}(k), \tag{3.42}$$

where

$$\boldsymbol{\Theta} = \begin{bmatrix} \boldsymbol{\Theta}_{u.l} \\ \boldsymbol{\Theta}_{y.l} \end{bmatrix}, \; \boldsymbol{z}_{uy.l}(k-1) = \begin{bmatrix} \boldsymbol{u}_l(k-1) \\ \boldsymbol{y}_l(k-1) \end{bmatrix}$$

with $\boldsymbol{\Theta}_{u.l}$ and $\boldsymbol{\Theta}_{y.l}$ being parameter matrices with proper dimensions. $\boldsymbol{u}_l(k-1)$ and $\boldsymbol{y}_l(k-1)$ are past I/O data of the plant $\boldsymbol{P}$ with length l, defined as

$$\boldsymbol{u}_l(k-1) = \begin{bmatrix} \boldsymbol{u}(k-1) \\ \boldsymbol{u}(k-2) \\ \vdots \\ \boldsymbol{u}(k-l) \end{bmatrix}, \; \boldsymbol{y}_l(k-1) = \begin{bmatrix} \boldsymbol{y}(k-1) \\ \boldsymbol{y}(k-2) \\ \vdots \\ \boldsymbol{y}(k-l) \end{bmatrix}, \tag{3.43}$$

and $\boldsymbol{\xi}(k) \sim N(\mathbf{0}, \boldsymbol{\Sigma}_{\xi})$ is Gaussian white noise that is used for persistent excitation. We vectorize $\boldsymbol{\Theta}$ with $\boldsymbol{\theta} = \text{vec}(\boldsymbol{\Theta})$, such that the optimization of $\boldsymbol{\Theta}$ can be performed on $\boldsymbol{\theta}$. To be consistent with the notation of existing studies on policy gradient methods in RL literature, we parameterize the control policy of $\boldsymbol{K}$ as

$$\begin{aligned}&\pi_{\theta}\big(\boldsymbol{u}(k)|z_{uy.l}(k-1)\big)\\&=\frac{1}{\sqrt{2\pi|\boldsymbol{\Sigma}_{\xi}|}}e^{-\frac{1}{2}\left\{\left(\boldsymbol{u}(k)-\boldsymbol{\Theta}^{T}z_{uy.l}(k-1)\right)^{T}\boldsymbol{\Sigma}_{\xi}^{-1}\left(\boldsymbol{u}(k)-\boldsymbol{\Theta}^{T}z_{uy.l}(k-1)\right)\right\}}.\end{aligned} \tag{3.44}$$

Here the stochastic policy $\pi_{\theta}\big(\boldsymbol{u}(k)|z_{uy.l}(k-1)\big)$ denotes the probability of taking the control $\boldsymbol{u}(k)$ conditioned on the current collected I/O data $z_{uy.l}(k-1)$.

(B). Parameterization of the Value Function and *Q*-function
Prior to the parameterization, we investigate the form of the value function that can evaluate performance of all output feedback controllers parameterized by (3.42).

As the plant $\boldsymbol{P}$ is unknown, to represent its dynamics precisely, the following Kalman I/O data model is used [56]

$$\begin{cases}\boldsymbol{x}_K(k+1) = \boldsymbol{A}\boldsymbol{x}_K(k) + \boldsymbol{B}\boldsymbol{u}(k) + \boldsymbol{L}_K\boldsymbol{r}_K(k),\\ \boldsymbol{r}_K(k) = \boldsymbol{y}(k) - \boldsymbol{C}\boldsymbol{x}_K(k) - \boldsymbol{D}\boldsymbol{u}(k),\end{cases} \tag{3.45}$$

where $\boldsymbol{x}_K$ is the Kalman state estimate and $\boldsymbol{L}_K$ is the Kalman gain, and $\boldsymbol{r}_K$ is the innovation signal satisfying $\boldsymbol{r}_K \sim N(\mathbf{0}, \boldsymbol{\Sigma}_r)$. Then it is evident that the state $\boldsymbol{x}_K$ can be equivalently represented by

$$\boldsymbol{x}_K(k) = \boldsymbol{A}_{L_K}^{h}\boldsymbol{x}_K(k-h) + \sum_{i=1}^{h}\boldsymbol{A}_{L_K}^{h-1}\left[\boldsymbol{B}_{L_K}\ \boldsymbol{L}_K\right]\begin{bmatrix}\boldsymbol{u}(k-1)\\ \boldsymbol{y}(k-1)\end{bmatrix}$$

where h is a positive integer, and $\boldsymbol{A}_{L_K} = \boldsymbol{A} - \boldsymbol{L}_K\boldsymbol{C}$ and $\boldsymbol{B}_{L_K} = \boldsymbol{B} - \boldsymbol{L}_K\boldsymbol{D}$. Considering that h is sufficiently large (typically much larger than the order n of the plant $\boldsymbol{P}$), $\boldsymbol{A}_{L_K}^{h} \approx \mathbf{0}$ holds and $\boldsymbol{x}_K(k)$ can be approximated by I/O data of $\boldsymbol{P}$ by [56]

$$\boldsymbol{x}_K(k) \approx \boldsymbol{M}_{uy}^{T}z_{uy.h}(k-1), \tag{3.46}$$

where $\boldsymbol{M}_{uy}$ is a mapping with $\boldsymbol{M}_{uy}^{T} = \left[\boldsymbol{M}_u\ \boldsymbol{M}_y\right]$,
$\boldsymbol{M}_u = \left[\boldsymbol{A}_{L_K}^{h-1}\boldsymbol{B}_{L_K}\ \cdots\ \boldsymbol{A}_{L_K}\boldsymbol{B}_{L_K}\ \boldsymbol{B}_{L_K}\right]$, $\boldsymbol{M}_y = \left[\boldsymbol{A}_{L_K}^{h-1}\boldsymbol{L}_K\ \cdots\ \boldsymbol{A}_{L_K}\boldsymbol{L}_K\ \boldsymbol{L}_K\right]$.
Here we choose a large length h satisfying $h \geq l$.

Theorem 3.1 *A value function, defined in (3.33), that evaluates all stabilizing output feedback controllers (3.42) of* $\boldsymbol{P}$ *(3.30) can be approximated by*

$$V\left(z_{uy.h}(k-1)\right) = z_{uy.h}^T(k-1)\,\boldsymbol{P_o} z_{uy.h}(k-1) + c_o, \tag{3.47}$$

where $\boldsymbol{P_o}$ *is a constant matrix, and* c_o *is a constant.*

Proof Initially, considering (3.45), (3.46), the value at the final time instant $N (N \to \infty)$ is [71]

$$\begin{aligned} V\left(z_{uy.h}(N-1)\right) &= \mathbb{E}\left\{\boldsymbol{u}^T(N)\,\boldsymbol{R}\boldsymbol{u}(N) + \boldsymbol{y}^T(N)\,\boldsymbol{Q}\boldsymbol{y}(N)\right\} \\ &= z_{uy.h}^T(N-1)\,\boldsymbol{P_o}(N)\, z_{uy.h}(N-1) + c_o(N), \end{aligned} \tag{3.48}$$

where

$$\boldsymbol{P_o}(N) = \boldsymbol{K}_1 = \begin{bmatrix} \boldsymbol{M}_{uy}^T \\ \boldsymbol{\Theta}^T \end{bmatrix}^T \begin{bmatrix} \boldsymbol{C}^T\boldsymbol{Q}\boldsymbol{C} & \boldsymbol{C}^T\boldsymbol{Q}\boldsymbol{D} \\ \boldsymbol{D}^T\boldsymbol{Q}\boldsymbol{C} & \boldsymbol{R} + \boldsymbol{D}^T\boldsymbol{Q}\boldsymbol{D} \end{bmatrix} \begin{bmatrix} \boldsymbol{M}_{uy}^T \\ \boldsymbol{\Theta}^T \end{bmatrix}, \; c_o(N) = \operatorname{tr}\left(\boldsymbol{\Sigma_r}\boldsymbol{Q}\right).$$

Next, we determine $V(N-1)$ and assume $l = h$ without loss of generality. The expectation of the one-step cost at $N-1$ can be represented by

$$\begin{aligned} &\mathbb{E}\left\{\boldsymbol{u}^T(N-1)\,\boldsymbol{R}\boldsymbol{u}(N-1) + \boldsymbol{y}^T(N-1)\,\boldsymbol{Q}\boldsymbol{y}(N-1)\right\} \\ &= z_{uy.h}^T(N-2)\,\boldsymbol{K}_1 z_{uy.h}(N-2) + \operatorname{tr}\left\{\boldsymbol{\Sigma_r}\boldsymbol{Q}\right\}, \end{aligned} \tag{3.49}$$

Considering (3.42), (3.45) and the following equalities

$$u_{h-1}(N-2) = \begin{bmatrix} \boldsymbol{I}_{g_1} \\ \boldsymbol{0}_{k_u \times g_1} \\ \boldsymbol{0}_{mh \times g_1} \end{bmatrix}^T \begin{bmatrix} u_{h-1}(N-2) \\ \boldsymbol{u}(N-h-1) \\ y_h(N-2) \end{bmatrix} = \boldsymbol{T_{c1}} z_{uy.h}(N-2),$$

$$y_{h-1}(N-2) = \begin{bmatrix} \boldsymbol{0}_{k_u h \times g_2} \\ \boldsymbol{I}_{g_2} \\ \boldsymbol{0}_{m \times g_2} \end{bmatrix}^T \begin{bmatrix} u_h(N-2) \\ y_{h-1}(N-2) \\ y(N-h-1) \end{bmatrix} = \boldsymbol{T_{c2}} z_{uy.h}(N-2),$$

where

$$\boldsymbol{T_{c1}} = \left[\boldsymbol{I}_{g_1}\; \boldsymbol{0}_{g_1 \times k_u}\; \boldsymbol{0}_{g_1 \times mh}\right], \boldsymbol{T_{c2}} = \left[\boldsymbol{0}_{g_2 \times k_u h}\; \boldsymbol{I}_{g_2}\; \boldsymbol{0}_{g_2 \times m}\right],$$

and $g_1 = k_u(h-1)$ and $g_2 = m(h-1)$, the relationship between $z_{uy.h}(N-1)$ and $z_{uy.h}(N-2)$ can be derived

$$z_{uy.h}(N-1) = \begin{bmatrix} u(N-1) \\ u_{h-1}(N-2) \\ y(N-1) \\ y_{h-1}(N-2) \end{bmatrix} = \begin{bmatrix} K_2 & K_3 \end{bmatrix} \begin{bmatrix} z_{uy.h}(N-2) \\ r_K(N-1) \end{bmatrix}, \tag{3.50}$$

where

$$K_2 = \begin{bmatrix} \Theta\ T_{c1}^T\ M_{uy}C^T + \Theta D^T\ T_{c2}^T \end{bmatrix}^T, K_3 = \begin{bmatrix} 0\ 0\ I_m\ 0 \end{bmatrix}^T.$$

Substituting (3.50) into (3.48) and considering (3.49), $V(N-1)$ can be derived as

$$\begin{aligned} V(N-1) &= \mathbb{E}\left(u^T(N-1)\,Ru(N-1) + y^T(N-1)\,Qy(N-1) + \gamma V(N)\right) \\ &= z_{uy.h}^T(N-2)\,P_o(N-1)\,z_{uy.h}(N-2) + c_o(N-1), \end{aligned} \tag{3.51}$$

where

$$\begin{aligned} P_o(N-1) &= K_1 + \gamma K_2^T P_o(N) K_2, \\ c_o(N-1) &= (1+\gamma)\,\mathrm{tr}(\Sigma_r Q) + \gamma \mathrm{tr}\left(\Sigma_r K_3^T P_o(N) K_3\right). \end{aligned}$$

Finally, repeat the above step to calculate values from the time instant $N-2$ backwards until convergence is reached. Inferred from (3.48), (3.51), the converged value function has the form (3.47). Meanwhile the converged P_o solves the following Lyapunov equation

$$P_o = K_1 + \gamma K_2^T P_o K_2,$$

and the constant c_o is [32]

$$c_o = \frac{\mathrm{tr}(\Sigma_r Q) + \gamma \mathrm{tr}\left(\Sigma_r K_3^T P_o K_3\right)}{1-\gamma}.$$

This gives Theorem 3.1. □

According to Theorem 3.1, we parameterize the value function, under the stochastic policy π_θ (3.44), by

$$V^{\pi_\theta}\left(z_{uy.h}(k-1)\right) = \left[\left(z_{uy.h}(k-1) \otimes z_{uy.h}(k-1)\right)^T\ 1\right] p_\theta, \tag{3.52}$$

where $\boldsymbol{p}_{\theta}$ is a parameterized column vector. Then the Q-function at k, when a control $\boldsymbol{u}(k)$ is taken, can be described by [55]

$$Q^{\pi_\theta}\left(z_{uy.h}\left(k-1\right),\boldsymbol{u}\left(k\right)\right)=\boldsymbol{u}^T\left(k\right)\boldsymbol{R}\boldsymbol{u}\left(k\right)+\boldsymbol{y}^T\left(k\right)\boldsymbol{Q}\boldsymbol{y}\left(k\right)+\gamma V^{\pi_\theta}\left(z_{uy.h}\left(k\right)\right) \tag{3.53}$$

(C). Policy Gradient Methods

The basic idea of policy gradient methods is to update the policy parameters according to the following gradient update rule

$$\boldsymbol{\theta}_{j+1}=\boldsymbol{\theta}_j-\alpha\nabla_{\theta}\,V^{\pi_\theta}\big|_{\theta=\theta_j}, \tag{3.54}$$

where $\nabla_\theta V^{\pi_\theta}$ is a shorthand for $\nabla_\theta V^{\pi_\theta}\left(z_{uy.h}\left(k-1\right)\right)$, and α is a positive step-size. The challenging problem is to find out an unbiased estimation of $\nabla_\theta V^{\pi_\theta}$. Fortunately, the policy gradient theorem [66] gives a theoretical answer, that is

$$\begin{aligned}\nabla_\theta V^{\pi_\theta}&=\int_{\mathbb{T}}p_\theta\left(\tau\right)\left(\sum_{i=k}^{\infty}\gamma^{i-k}Q^{\pi_\theta}\left(z_{uy.h}\left(i-1\right),\boldsymbol{u}\left(i\right)\right)\nabla_\theta\ln\pi_\theta\left(\boldsymbol{u}(i)|z_{uy.l}(i-1)\right)\right)d\tau\\&=\mathbb{E}_\tau\left(\sum_{i=k}^{\infty}\gamma^{i-k}Q^{\pi_\theta}\left(z_{uy.h}\left(i-1\right),\boldsymbol{u}\left(i\right)\right)\nabla_\theta\ln\pi_\theta\left(\boldsymbol{u}(i)|z_{uy.l}(i-1)\right)\right),\end{aligned} \tag{3.55}$$

where $\tau=\left\{z_{uy.h}\left(k-1\right),\boldsymbol{u}\left(k\right),z_{uy.h}\left(k\right),\boldsymbol{u}(k+1),\cdots\right\}$ represents a system trajectory by taking controls sampled from π_θ from the time instant k onward, and its probability distribution is $p_\theta(\tau)$, and the set of all trajectories is denoted by $\mathbb{T}$. $\mathbb{E}_\tau$ denotes the expectation is taken over all possible system trajectories. It can be seen that Equation (3.55) converts the complex computation of the policy gradient to a simple expectation. Based on this, the parameter vector $\boldsymbol{\theta}$ can be updated with SGD methods. However, the acquirement of the Q-function is still an issue. As it is unknown, an approximation is required. This can lead to high variance of the estimated policy gradient. It is given in [58, 65] that a value function baseline can be introduced to the policy gradient estimation (3.55) to reduce the variance, which results in

$$\nabla_\theta V^{\pi_\theta}=\mathbb{E}_\tau\left(\sum_{i=k}^{\infty}\gamma^{i-k}A^{\pi_\theta}\left(z_{uy.h}\left(i-1\right),\boldsymbol{u}\left(i\right)\right)\nabla_\theta\ln\pi_\theta\left(\boldsymbol{u}(i)|z_{uy.l}(i-1)\right)\right), \tag{3.56}$$

$$A^{\pi_\theta}\left(z_{uy.h}\left(i-1\right),\boldsymbol{u}\left(i\right)\right)=Q^{\pi_\theta}\left(z_{uy.h}\left(i-1\right),\boldsymbol{u}\left(i\right)\right)-V^{\pi_\theta}\left(z_{uy.h}\left(i-1\right)\right), \tag{3.57}$$

where A^{π_θ} is called advantage function, representing the expected more cost incurred by taking the control input $\boldsymbol{u}(k)$ than by taking all possible inputs sampled from the policy π_θ when the past data $z_{uy.l}(k-1)$ is collected. However, the key issue in calculating the policy gradient still persists, which now becomes the acquirement of A^{π_θ}. To solve this, an actor-critic architecture is applied, in which an actor is used for the parameterization of the control policy π_θ, and a critic is used as a function approximator $A^{\boldsymbol{w}_\theta}$ for the estimation of the true advantage function A^{π_θ} with $\boldsymbol{w}_\theta$ being the parameter vector.

To avoid biases in estimation of both $\nabla_\theta V^{\pi_\theta}$ and A^{π_θ} due to the use of the actor-critic structure, it is required that $A^{\boldsymbol{w}_\theta}$ should be compatible [66]. The following two conditions should be satisfied: i)

$$A^{\pi_\theta}\left(z_{uy.h}(k-1), \boldsymbol{u}(k)\right) = \left(\nabla_\theta \ln \pi_\theta\left(\boldsymbol{u}(k)|z_{uy.l}(k-1)\right)\right)^T \boldsymbol{w}_\theta, \tag{3.58}$$

which guarantees no bias in the estimation of $\nabla_\theta V^{\pi_\theta}$. ii) The parameter vector $\boldsymbol{w}_\theta$ should be chosen to minimize the mean-squared error ε between A^{π_θ} and $A^{\boldsymbol{w}_\theta}$

$$\varepsilon = \mathbb{E}_\tau\left(\sum_{i=k}^{\infty}\gamma^{i-k}\left(A^{\pi_\theta}\left(z_{uy.h}(i-1), \boldsymbol{u}(i)\right) - A^{\boldsymbol{w}_\theta}\left(z_{uy.h}(i-1), \boldsymbol{u}(i)\right)\right)^2\right). \tag{3.59}$$

It is worth mentioning that the second condition is often relaxed when TD learning is used [63].

Substituting (3.58) into (3.56) and recomputing $\nabla_\theta V^{\pi_\theta}$, it is derived as [54]

$$\nabla_\theta V^{\pi_\theta} = \mathbb{E}_\tau\left(\begin{array}{l}\sum_{i=k}^{\infty}\gamma^{i-k}\nabla_\theta \ln \pi_\theta\left(\boldsymbol{u}(i)|z_{uy.l}(i-1)\right) \\ \times\left(\nabla_\theta \ln \pi_\theta\left(\boldsymbol{u}(i)|z_{uy.l}(i-1)\right)\right)^T\end{array}\right)\boldsymbol{w}_\theta = \boldsymbol{F}_\theta \boldsymbol{w}_\theta, \tag{3.60}$$

where

$$\boldsymbol{F}_\theta = \mathbb{E}_\tau\left(\sum_{i=k}^{\infty}\gamma^{i-k}\nabla_\theta \ln \pi_\theta\left(\boldsymbol{u}(i)|z_{uy.l}(i-1)\right)\left(\nabla_\theta \ln \pi_\theta\left(\boldsymbol{u}(i)|z_{uy.l}(i-1)\right)\right)^T\right). \tag{3.61}$$

It is clear that one must identify both $\boldsymbol{F}_\theta$ and $\boldsymbol{w}_\theta$ in order to compute the policy gradient.

(D). Natural Policy Gradient

The policy update rule given in (3.54) can be regarded as updating the policy parameter vector $\boldsymbol{\theta}$ in the steepest descent direction of $\nabla_{\boldsymbol{\theta}} V^{\pi_{\boldsymbol{\theta}}}$ by $\Delta\boldsymbol{\theta}$ under the constraint that the squared length $\Delta\boldsymbol{\theta}^T \Delta\boldsymbol{\theta}$ is held constant. One critical problem with the policy update is that it is not affine invariant [2], namely, if there exists a linear coordinate transformation $\boldsymbol{h} = \boldsymbol{T}^{-1}\boldsymbol{\theta}$ and $\boldsymbol{h}_j = \boldsymbol{T}^{-1}\boldsymbol{\theta}_j$, then $\boldsymbol{h}_{j+1} \neq \boldsymbol{T}^{-1}\boldsymbol{\theta}_{j+1}$ using (3.54). This can easily lead to slow learning. In order to deal with it, Kakade [37] introduced a natural gradient [2] into RL. The natural gradient uses, instead of the squared length, the distance between the probability distribution of the trajectory of the existing policy $p_{\boldsymbol{\theta}}(\tau)$ and that of the updated policy $p_{\boldsymbol{\theta}+\Delta\boldsymbol{\theta}}(\tau)$, as the constraint of $\Delta\boldsymbol{\theta}$. A commonly used distance measure is the Kullback-Leibler (KL) divergence $d_{\mathrm{KL}}\left(p_{\boldsymbol{\theta}}(\tau) \parallel p_{\boldsymbol{\theta}+\Delta\boldsymbol{\theta}}(\tau)\right)$. If $\Delta\boldsymbol{\theta}$ is sufficiently small, then the KL divergence can be approximated by the second order Taylor expansion by [53]

$$d_{\mathrm{KL}}\left(p_{\boldsymbol{\theta}}(\tau) \parallel p_{\boldsymbol{\theta}+\Delta\boldsymbol{\theta}}(\tau)\right) \approx \Delta\boldsymbol{\theta}^T \boldsymbol{G}_{\boldsymbol{\theta}} \Delta\boldsymbol{\theta}, \tag{3.62}$$

where $\boldsymbol{G}_{\boldsymbol{\theta}} = \boldsymbol{F}_{\boldsymbol{\theta}}$ is Fisher information matrix. Considering (3.60) and the constraint that the KL divergence $d_{\mathrm{KL}}\left(p_{\boldsymbol{\theta}}(\tau) \parallel p_{\boldsymbol{\theta}+\Delta\boldsymbol{\theta}}(\tau)\right)$ (3.62) is held constant, the natural gradient is derived by

$$\tilde{\nabla}_{\boldsymbol{\theta}} V^{\pi_{\boldsymbol{\theta}}} = \boldsymbol{G}_{\boldsymbol{\theta}}^{-1} \nabla_{\boldsymbol{\theta}} V^{\pi_{\boldsymbol{\theta}}} = \boldsymbol{F}_{\boldsymbol{\theta}}^{-1} \nabla_{\boldsymbol{\theta}} V^{\pi_{\boldsymbol{\theta}}} = \boldsymbol{w}_{\boldsymbol{\theta}}. \tag{3.63}$$

Compared with $\nabla_{\boldsymbol{\theta}} V^{\pi_{\boldsymbol{\theta}}}$, the natural gradient has several advantages highlighted in [55]:

- It has faster convergence and avoids premature convergence from empirical observations, as it chooses a more direct path to the optimal solution in the parameter space.
- It is affine invariant, i.e., independent of the coordinate system used for expressing the policy parameter vector.
- It requires fewer data points for a good gradient estimate, as it allows the gradient estimate to be analytically computed based on the average influence of the stochastic policy.

NAC adopts the natural gradient and its policy update rule can be derived as

$$\boldsymbol{\theta}_{j+1} = \boldsymbol{\theta}_j - \alpha_n \tilde{\nabla}_{\boldsymbol{\theta}}\, V^{\pi_{\boldsymbol{\theta}}}\Big|_{\boldsymbol{\theta}=\boldsymbol{\theta}_j} = \boldsymbol{\theta}_j - \alpha_n \boldsymbol{w}_{\boldsymbol{\theta}_j}, \tag{3.64}$$

where α_n is a positive constant step-size.

(E). Data-driven Implementation of NAC
Seen from (3.64), the key to update the parameter vector $\boldsymbol{\theta}_j$ of the policy $\pi_{\boldsymbol{\theta}_j}$ is the identification of $\boldsymbol{w}_{\boldsymbol{\theta}_j}$. The logarithmic derivative of the stochastic policy (3.44) evaluated at $\boldsymbol{\theta}_j$ is

$$\begin{aligned}&\nabla_{\boldsymbol{\theta}} \ln \pi_{\boldsymbol{\theta}}\big(\boldsymbol{u}(k)|\boldsymbol{z}_{uy.l}(k-1)\big)\big|_{\boldsymbol{\theta}=\boldsymbol{\theta}_j}\\ &= \big(\boldsymbol{I}_{k_u} \otimes \boldsymbol{z}_{uy.l}(k-1)\big)\,\boldsymbol{\Sigma}_{\xi}^{-1}\left(\boldsymbol{u}(k) - \boldsymbol{\Theta}_j^T \boldsymbol{z}_{uy.l}(k-1)\right). \end{aligned} \tag{3.65}$$

Substituting (3.52), (3.53) and (3.58) into the corresponding items in (3.57), and considering (3.65), Equation (3.57) can be rewritten in the form of TD error that is

$$\begin{aligned}\delta(k)=&\left[\left(\boldsymbol{z}_{uy.h}(k-1)\otimes \boldsymbol{z}_{uy.h}(k-1)\right)^T\ 1\right]\boldsymbol{p}_{\boldsymbol{\theta}_j} - \boldsymbol{u}^T(k)\,\boldsymbol{R}\boldsymbol{u}(k) - \boldsymbol{y}^T(k)\,\boldsymbol{Q}\boldsymbol{y}(k)\\ &-\gamma\left[\left(\boldsymbol{z}_{uy.h}(k)\otimes \boldsymbol{z}_{uy.h}(k)\right)^T\ 1\right]\boldsymbol{p}_{\boldsymbol{\theta}_j}\\ &+\left(\boldsymbol{I}_{k_u}\otimes \boldsymbol{z}_{uy.l}(k-1)\right)\boldsymbol{\Sigma}_{\xi}^{-1}\left(\boldsymbol{u}(k)-\boldsymbol{\Theta}_j^T \boldsymbol{z}_{uy.l}(k-1)\right)^T \boldsymbol{w}_{\boldsymbol{\theta}_j}.\end{aligned} \tag{3.66}$$

Minimizing this error using LS or SGD, parameters $\boldsymbol{p}_{\boldsymbol{\theta}_j}$ and $\boldsymbol{w}_{\boldsymbol{\theta}_j}$ can be online identified with a sequence of $\boldsymbol{u}$ and $\boldsymbol{y}$ data collected under the stochastic policy $\pi_{\boldsymbol{\theta}_j}$. Subsequently, the policy parameter vector can be optimized by (3.64).

To sum up, the data-driven algorithm for optimizing the parameter vector $\boldsymbol{\theta}$ of the controller $\boldsymbol{K}$ is given in Algorithm 3.7.

Algorithm 3.7 On-policy NAC: Data-driven Optimization of the Controller $\boldsymbol{K}$

1: **Initialization**:

- Choose a sufficiently large length h of the past $\boldsymbol{u}$ and $\boldsymbol{y}$ data for the parameterization of the value function (3.52);
- Choose a length l ($l \leq h$) of past $\boldsymbol{u}$ and $\boldsymbol{y}$ data for the parameterization of the controller $\boldsymbol{K}$ (3.42);
- Select an initial control parameter vector $\boldsymbol{\theta}_0$ and a positive constant step-size α_n.

for $j = 0, 1, \ldots$
2: **Repeat**
3: **Policy evaluation**: Identify $\boldsymbol{p}_{\boldsymbol{\theta}_j}$ and $\boldsymbol{w}_{\boldsymbol{\theta}_j}$ using LS or SGD by minimizing the TD error (3.66) with a sequence of $\boldsymbol{u}$ and $\boldsymbol{y}$, and control parameter $\boldsymbol{\theta}_j$.
4: **Policy improvement**: Update $\boldsymbol{\theta}_j$ to $\boldsymbol{\theta}_{j+1}$ using (3.64).
5: **Until** $\boldsymbol{\theta}$ converges or some end conditions are satisfied.

Although, in practice, the LQG controller is hardly achievable in a data-driven manner due to the stochastic noise in the system, the on-policy NAC method provides an effective means to learn an optimal output feedback controller with a prescribed controller structure. Often, a low-order output feedback controller is preferred for online implementation, as fewer online computation is involved. In Chapter 5, we will further investigate more efficient NAC methods for performance optimization using both plant model and data.

3.3.3 Simulation Results

In this subsection, the proposed new NAC learning method is applied to learn an optimal output feedback controller with a prescribed structure for a multiple-input multiple-output (MIMO) plant. The plant is described in the following continuous-time state-space form

$$\dot{x} = Ax + B(u + \omega), \quad y = Cx + D(u + \omega) + v, \tag{3.67}$$

$$A = \begin{bmatrix} -167.723 & 2001 & 408.688 \\ -1976 & -223.083 & -2163 \\ -516.023 & 2139 & -176.143 \end{bmatrix}, \; B = \begin{bmatrix} 83.279 & -1693 \\ -270.807 & 444.997 \\ -300.602 & -2160 \end{bmatrix},$$

$$C = \begin{bmatrix} -0.384 & -0.109 & 0 \\ 0 & 0.182 & -0.851 \end{bmatrix}, \qquad D = \begin{bmatrix} 0 & 0.515 \\ 0 & 0 \end{bmatrix},$$

$$\omega \sim N\left(\begin{bmatrix} 0 \\ 0 \end{bmatrix}, \begin{bmatrix} 10^{-3} & 0 \\ 0 & 10^{-3} \end{bmatrix}\right), \qquad v \sim N\left(\begin{bmatrix} 0 \\ 0 \end{bmatrix}, \begin{bmatrix} 10^{-4} & 0 \\ 0 & 10^{-4} \end{bmatrix}\right).$$

The cost function is defined in the form of (3.33) with the following weighting matrices

$$R = \begin{bmatrix} 0.01 & 0 \\ 0 & 0.01 \end{bmatrix}, \; Q = \begin{bmatrix} 50 & 20 \\ 20 & 10 \end{bmatrix},$$

and the discount factor $\gamma = 0.99$ and the sampling time is 1 ms. Theoretically, the optimal state feedback gain F^* and the Kalman gain L^* of the LQG controller can be calculated based on Equation (3.10) and Equation (3.40). They are

$$F^* = \begin{bmatrix} -3.5627 & 0.2108 & -0.7436 \\ 0.7411 & 0.0708 & 0.6595 \end{bmatrix}, \; L^* = \begin{bmatrix} -0.0531 & 0.7045 \\ 2.3906 & -0.4309 \\ 0.0447 & -0.0494 \end{bmatrix}.$$

We parameterize the output feedback controller $\boldsymbol{K}$ in the form (3.42) with a length $l = 1$. To optimize $\boldsymbol{\Theta}$ of $\boldsymbol{K}$ with the proposed NAC method, we parameterize the value function in the form (3.52) with a length $h = 9$. Fig. 3.4 shows the changes of $\boldsymbol{\Theta}$ and the cost J over the number of policy updates during online learning. Note that the 'Cost' marking the y-axis of the lower plot of Fig. 3.4 measures the total cost within a time interval needed for one policy update. It is computed by

$$J(i) = \sum_{k=Ni}^{N(i+1)-1} \gamma^{(k-Ni)} \left(\boldsymbol{u}^T(k) \boldsymbol{R} \boldsymbol{u}(k) + \boldsymbol{y}^T(k) \boldsymbol{Q} \boldsymbol{y}(k) \right),$$

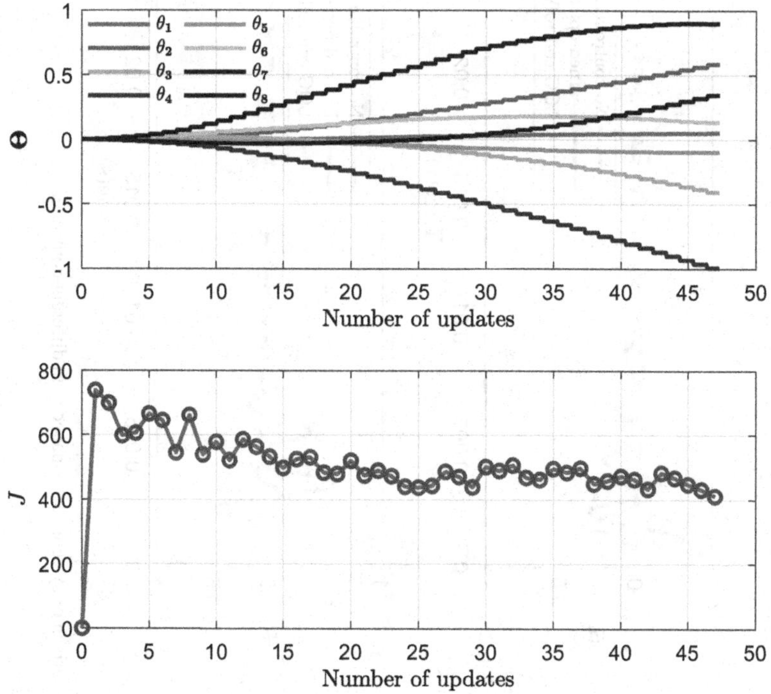

Figure 3.4 Parameters $\boldsymbol{\Theta}$ and cost J during online learning

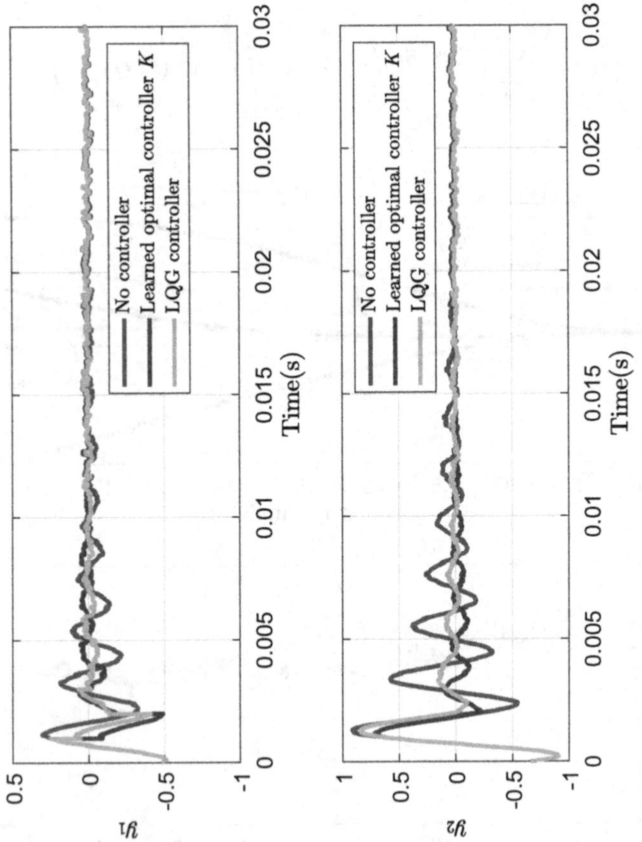

Figure 3.5 Comparison of system performance with different controllers

where N is chosen to be 1112 that is equal to 1.112 s, and i is an index denoting the ith policy update. Inferred from the gradual decreasing cost in Fig. 3.4, the updating directions of the parameters of $\boldsymbol{K}$ are correct.

Finally, after 48 policy updates, the learning procedure is stopped. The learned $\boldsymbol{K}$ with the parameter

$$\boldsymbol{\Theta} = \begin{bmatrix} 0.05141 & -0.0953 \\ 0.5842 & 0.1327 \\ -0.4054 & 0.8991 \\ -0.9925 & 0.3451 \end{bmatrix}$$

is placed into the system. Fig. 3.5 shows the comparison of system responses at the initial state $\boldsymbol{x}_0 = \begin{bmatrix} 1 & 1 & 1 \end{bmatrix}^T$ in three scenarios: open-loop control, closed-loop control with the learned controller $\boldsymbol{K}$, and closed-loop control with the LQG controller. It can be seen from Fig. 3.5 that, compared with the open-loop response, the response of the system with the learned controller $\boldsymbol{K}$ has considerably less oscillation and less settling time for both outputs. Furthermore, the performance of the system with $\boldsymbol{K}$ is very close to that of the system with the LQG controller. This demonstrates the effectiveness of the proposed NAC method for the optimization of an output feedback controller with a prescribed structure.

3.4 Concluding Remarks

In this chapter, we have briefly introduced RL and feedback control, and reviewed two fundamental feedback control optimization problems, LQR and LQG. In both cases, DP and RL methods are applied to obtain the optimal policies by minimizing prescribed value (or cost) functions.

The key idea of both DP and RL is the use of the Bellman equation and the Bellman optimality equation, respectively, for policy evaluation and policy improvement. Based on this, we obtain the two most popular DP-based methods, policy iteration and value iteration. They are applied to solve the LQR and LQG problems using DP.

To effectively solve the LQR problem using RL, we have introduced two critic-only methods, off-policy Q-learning and on-policy SARSA. Empirically, Q-learning shows superior performance over SARSA in terms of accuracy and speed of convergence. However, because their updates of a control policy are too aggressive and rely heavily on the precisely identified value functions, they cannot be applied to solve the LQG problem. To handle it, a new NAC method is proposed, which is capable of learning an optimal output feedback controller with a prescribed structure.

Q-learning Aided Performance Optimization of Deterministic Systems

4

Often, when adverse changes or perturbations happen in industrial systems, it is desirable to retain the original control structure and improve the capability of the existing controller to minimize the control performance degradation, rather than decommission the whole system and replace it with a new one. We aim to solve such a performance degradation optimization problem in a data-driven fashion in this and the subsequent chapter.

The main goal of this chapter is to provide two approaches to deal with performance degradation. To simplify our study, we consider deterministic feedback control systems where no randomness is involved. The chapter is organized as follows. First, we give a new input and output recovery method, which provides an alternative approach to many existing methods, such as loop transfer recovery, for the optimization of system robustness against perturbations. Then, we develop a performance index based method for the optimization of system performance. Finally, we generalize the latter solution to the optimization of system tracking performance. For data-driven implementation of the above developed approaches, we will use Q-learning throughout this chapter.

4.1 Problem Formulation

We deal with the plant $\boldsymbol{P_0}$ (2.1). Assume that an initial stabilizing observer-based state feedback controller $\boldsymbol{K_0}$ (2.4) is designed with a feedback gain $\boldsymbol{F}$ and an observer gain $\boldsymbol{L}$ guaranteeing good performance based on the performance index (3.2).

It can be envisioned that system performance would degrade over time due to several perturbations happen in the plant, such as changes in operation conditions, e.g. loading and unloading, or replacement of components in maintenance actions. Assume that the perturbations degrade only system performance and have no influ-

C. Hua, *Reinforcement Learning Aided Performance Optimization of Feedback Control Systems*, https://doi.org/10.1007/978-3-658-33034-7_4

ence on stability. According to Theorem 2.5, we denote the perturbed plant as $P(S_f)$ with the perturbations termed $S_f \in \mathcal{RH}_\infty$. The initial control structure is shown in Fig. 4.1, where s equals $\mathbf{0}$.

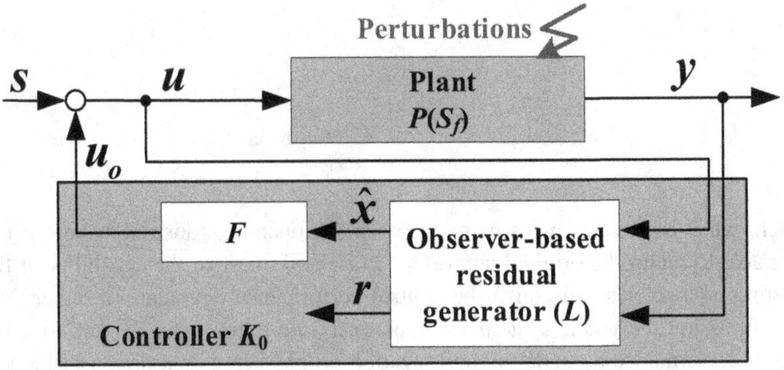

Figure 4.1 The initial control structure

In order to obtain practical solutions to performance degradation, we impose the following three conditions to controller reconfiguration mechanisms to be developed:

- they introduce no modifications to the existing controller;
- they can handle diverse plant changes, such as parametric or structural changes, as long as the system is linear;
- they have low online computational cost.

Note that these conditions are applied to all performance optimization methods to be developed throughout the thesis. In the next two sections, we will introduce two approaches to deal with performance degradation.

4.2 Robustness Optimization

Oftentimes, when it comes to optimization of controllers, system robustness and performance are two critical measures. Different techniques are often used for controller design, such as LQR, LQG (or $\mathcal{H}_2$), or $\mathcal{H}_\infty$ controller. The LQR is the most desirable design as it shows a nice performance in terms of a quadratic performance index and, meanwhile, a nice robust property with an infinite gain margin and at least 60° phase margin [57]. However, in practice, due to inaccessible system states, an observer-based state feedback controller is often adopted. In such a case, good system performance can be maintained by the design, such as LQG or $\mathcal{H}_2$, but the nice robust property does not exist any more [18].

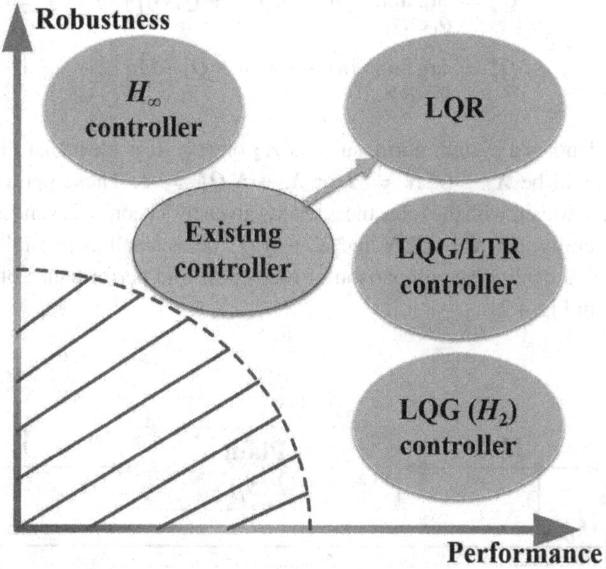

Figure 4.2 Comparison of robustness and performance of different controllers

4.2.1 Existing Robustness Optimization Approaches

In order to recover the good robust property of LQR, loop transfer recovery (LTR) techniques are widely applied. There are two main approaches well developed in the existing literature. One kind of LTR methods is to recover the loop gain transfer function of the nominal controller $\boldsymbol{K_0}$ to that of LQR by the redesign of the state feedback gain $\boldsymbol{F}$ or the observer gain $\boldsymbol{L}$ of $\boldsymbol{K_0}$ [4, 17]. However, this robustness recovery is obtained by increasing loop gains in the observer or state feedback design [50]. This results in, eventually, a high gain system that is sensitive to system noise, external disturbances or plant parameter changes. The other kind of LTR methods is to recover the input/output sensitivity transfer functions of LQG to that of LQR [50, 68]. This is achieved by the design of the YK parameter $\boldsymbol{Q_r}$ (2.17), which feeds back the residual signal to the plant input. Mathematically, the problem can be formulated, by considering the LCF and RCF of $\boldsymbol{P_0}$ and $\boldsymbol{K_0}$ in (2.7) and (2.8), as finding the optimal $\boldsymbol{Q_r^*}$ that has

$$\boldsymbol{Q_r^*} = \arg\min_{\boldsymbol{Q_r} \in \mathcal{RH}_\infty} \left\| \boldsymbol{M} - \boldsymbol{M}\left(\boldsymbol{X} - \boldsymbol{Q_r}\hat{\boldsymbol{N}}\right) \right\| \text{ or}$$

$$\boldsymbol{Q_r^*} = \arg\min_{\boldsymbol{Q_r} \in \mathcal{RH}_\infty} \left\| \hat{\boldsymbol{M}} - \left(\hat{\boldsymbol{X}} + \boldsymbol{N}\boldsymbol{Q_r}\right)\hat{\boldsymbol{M}} \right\|,$$

where $\|\cdot\|$ denotes a system norm, such as $\mathcal{H}_2$ or $\mathcal{H}_\infty$. It is clear that the optimal solution should be $\boldsymbol{X} - \boldsymbol{Q_r^*}\hat{\boldsymbol{N}} = \boldsymbol{I}$ or $\hat{\boldsymbol{X}} + \boldsymbol{N}\boldsymbol{Q_r^*} = \boldsymbol{I}$. These optimal results contradict, however, with the robustness results given by Theorem 2.3 and 2.4, which require generally that $\boldsymbol{X} - \boldsymbol{Q_r^*}\hat{\boldsymbol{N}}$ and $\hat{\boldsymbol{X}} + \boldsymbol{N}\boldsymbol{Q_r^*}$ be as small as possible. To sum up, we show a qualitative comparison of robustness and performance of different controllers in Fig. 4.2.

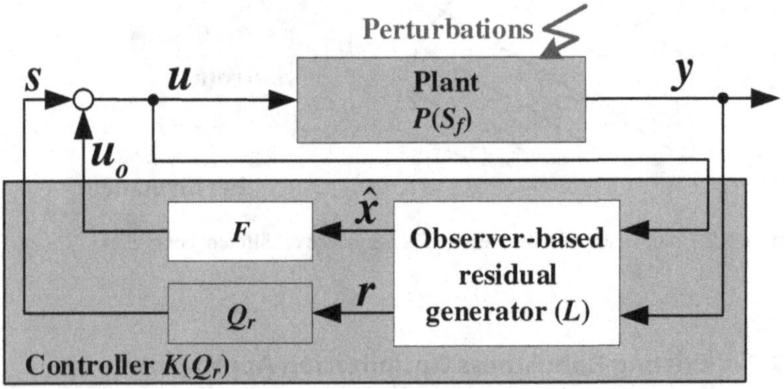

Figure 4.3 The performance optimization structure with $\boldsymbol{Q_r}$

Motivated by the foregoing studies, in this section, we propose a new input and output recovery method that enables the robustness of an observer-based controller to be gracefully recovered to that of LQR by augmenting the original observer-based controller $\boldsymbol{K_0}$ with an additional filter $\boldsymbol{Q_r}$. It is illustrated in Fig. 4.3. In addition, it is implemented by Q-learning in a data-driven manner.

4.2.2 Input and Output Recovery

The idea of input and output recovery (IOR) [15] is to minimize the discrepancy between the input/output (I/O) responses of an ideal system consisting of the nominal plant $\boldsymbol{P}_0$ and an LQR controller, and those of the actual system consisting of $\boldsymbol{P}(\boldsymbol{S}_f)$ and $\boldsymbol{K}(\boldsymbol{Q}_r)$. We first formulate the ideal closed-loop dynamics by

$$\begin{cases} \boldsymbol{x}_{ideal}(k+1) = \boldsymbol{A}\boldsymbol{x}_{ideal}(k) + \boldsymbol{B}\boldsymbol{u}_{ideal}(k), \\ \boldsymbol{y}_{ideal}(k) = \boldsymbol{C}\boldsymbol{x}_{ideal}(k) + \boldsymbol{D}\boldsymbol{u}_{ideal}(k), \\ \boldsymbol{u}_{ideal}(k) = \boldsymbol{F}\boldsymbol{x}_{ideal}(k), \end{cases} \tag{4.1}$$

where $\boldsymbol{x}_{ideal}$ is the ideal plant states. $\boldsymbol{F}$ is the designed optimal state feedback gain based on the nominal plant model. Next, the I/O response of the actual system is described using the observer I/O data model, by

$$\begin{cases} \hat{\boldsymbol{x}}(k+1) = \boldsymbol{A}\hat{\boldsymbol{x}}(k) + \boldsymbol{B}\boldsymbol{u}(k) + \boldsymbol{L}\boldsymbol{r}(k), \\ \boldsymbol{y}(k) = \boldsymbol{C}\hat{\boldsymbol{x}}(k) + \boldsymbol{D}\boldsymbol{u}(k) + \boldsymbol{r}(k), \\ \boldsymbol{u}(k) = \boldsymbol{F}\hat{\boldsymbol{x}}(k) + \boldsymbol{Q}_r\boldsymbol{r}(k). \end{cases} \tag{4.2}$$

The dynamics of the I/O differences can be readily obtained by subtracting (4.1) from (4.2), that is

$$\begin{aligned} \boldsymbol{e}_x(k+1) &= (\boldsymbol{A} + \boldsymbol{B}\boldsymbol{F})\boldsymbol{e}_x(k) + (\boldsymbol{B}\boldsymbol{Q}_r + \boldsymbol{L})\boldsymbol{r}(k), \\ \boldsymbol{e}_u(k) &= \boldsymbol{F}\boldsymbol{e}_x(k) + \boldsymbol{Q}_r\boldsymbol{r}(k), \\ \boldsymbol{e}_y(k) &= (\boldsymbol{C} + \boldsymbol{D}\boldsymbol{F})\boldsymbol{e}_x(k) + (\boldsymbol{D}\boldsymbol{Q}_r + \boldsymbol{I})\boldsymbol{r}(k), \end{aligned} \tag{4.3}$$

where $\boldsymbol{e}_x = \hat{\boldsymbol{x}} - \boldsymbol{x}_{ideal}$ is the state difference, and $\boldsymbol{e}_u = \boldsymbol{u} - \boldsymbol{u}_{ideal}$ and $\boldsymbol{e}_y = \boldsymbol{y} - \boldsymbol{y}_{ideal}$ denote I/O differences. Considering the LCF and RCF of the nominal plant and controller, the dynamics of the I/O differences can be equivalently written using z-domain transformation by

$$\begin{bmatrix} \boldsymbol{e}_u(z) \\ \boldsymbol{e}_y(z) \end{bmatrix} = \begin{bmatrix} -\hat{\boldsymbol{Y}}(z) + \boldsymbol{M}(z)\boldsymbol{Q}_r(z) \\ \hat{\boldsymbol{X}}(z) + \boldsymbol{N}(z)\boldsymbol{Q}_r(z) \end{bmatrix} \boldsymbol{r}(z). \tag{4.4}$$

It can be analyzed that the L_2 norm of the I/O differences expressed in (4.4) is upper-bounded

$$\min \left\| \begin{bmatrix} e_u \\ e_y \end{bmatrix} \right\|_2 = \min \left\| \begin{bmatrix} -\hat{Y} + MQ_r \\ \hat{X} + NQ_r \end{bmatrix} r \right\|_2$$
$$= \min \left\| \begin{bmatrix} M & -\hat{Y} \\ N & \hat{X} \end{bmatrix} \begin{bmatrix} s \\ r \end{bmatrix} \right\|_2 \leq \left\| \begin{bmatrix} M & -\hat{Y} \\ N & \hat{X} \end{bmatrix} \right\|_\infty \min \left\| \begin{bmatrix} s \\ r \end{bmatrix} \right\|_2 . \quad (4.5)$$

Remark 4.1 *The minimization of I/O differences (4.5) to improve robustness of the perturbed system is consistent with the robustness result given by the Theorem 2.3. It is also clear that an initial robust nominal controller $\boldsymbol{K}_0$ can be designed by choosing $\boldsymbol{F}$ and $\boldsymbol{L}$ to minimize* $\left\| \begin{bmatrix} M & -\hat{Y} \\ N & \hat{X} \end{bmatrix} \right\|_\infty$.

It can be concluded from (4.5) that if the upper bound of the L_2 norm of the I/O differences is minimized, then the L_2 norm of the differences themselves can also be minimized. Note that this is only a suboptimal design method. Even so, this design approach can be well accepted as $\boldsymbol{e}_u$ and $\boldsymbol{e}_y$ are immeasurable. Accordingly, for the design of $\boldsymbol{Q}_r$, we define a performance index by the following value function

$$V(k) = \lim_{N\to\infty} \sum_{i=k}^{N} \gamma^{i-k} \big(\boldsymbol{s}^T(i) \boldsymbol{W}_s \boldsymbol{s}(i) + \boldsymbol{r}^T(i) \boldsymbol{W}_r \boldsymbol{r}(i) \big), \quad (4.6)$$

where $\boldsymbol{W}_s > \boldsymbol{0}$ and $\boldsymbol{W}_r \geq \boldsymbol{0}$ are weighting matrices for the compensation signal $\boldsymbol{s}$ and the residual signal $\boldsymbol{r}$, respectively.

Remark 4.2 *Optimizing $\boldsymbol{Q}_r$ by minimizing the defined value function ensures the stability of the $\boldsymbol{S}_f$ and $\boldsymbol{Q}_r$ pair. Therefore, the perturbed plant $\boldsymbol{P}(\boldsymbol{S}_f)$ and controller $\boldsymbol{K}(\boldsymbol{Q}_r)$ is also stable according to Theorem 2.6.*

4.2.3 Robustness Optimization Using Q-learning

In this section, we propose a Q-learning algorithm for the tuning of $\boldsymbol{Q}_r$ to optimize system robustness against perturbations.

As $\boldsymbol{S}_f$ is unknown and no noise is considered, we represent it in the following state-space form

$$\boldsymbol{S}_f : \begin{cases} \boldsymbol{x}_s(k+1) = \boldsymbol{A}_s \boldsymbol{x}_s(k) + \boldsymbol{B}_s \boldsymbol{s}(k), \\ \boldsymbol{r}(k) = \boldsymbol{C}_s \boldsymbol{x}_s(k) + \boldsymbol{D}_s \boldsymbol{s}(k), \end{cases} \quad (4.7)$$

where $x_s \in \mathbb{R}^{k_s}$ is the state vector, A_s, B_s, C_s, D_s are unknown constant matrices with proper dimensions. Assume that S_f is both controllable and observable. Add and subtract the term $L_d r(k)$ to the right-hand side of the state equation in (4.7). It is evident that the state x_s can be equivalently represented by

$$x_s(k) = A_{L_d}^h x_s(k-h) + \sum_{i=1}^{h} A_{L_d}^{h-1} \left[B_{L_d} \; L_d \right] \begin{bmatrix} s(k-1) \\ r(k-1) \end{bmatrix}$$

where h is a positive integer, and $A_{L_d} = A_s - L_d C_s$ and $B_{L_d} = B_s - L_d D_s$. Considering that L_d is a deadbeat observer gain, we have $\forall h \geq k_s$, $A_{L_d}^h = \mathbf{0}$ and

$$x_s(k) = M_{sr}^T z_{sr.h}(k-1), \forall k \geq k_s, \tag{4.8}$$

where M_{sr} is a mapping with $M_{sr}^T = \left[M_s \; M_r \right]$,
$M_s = \left[A_{L_d}^{h-1} B_{L_d} \cdots A_{L_d} B_{L_d} \; B_{L_d} \right]$, $M_r = \left[A_{L_d}^{h-1} L_d \cdots A_{L_d} L_d \; L_d \right]$, and

$$z_{sr.h}(k-1) = \begin{bmatrix} s_h(k-1) \\ r_h(k-1) \end{bmatrix}, \; s_h(k-1) = \begin{bmatrix} s(k-1) \\ s(k-2) \\ \vdots \\ s(k-h) \end{bmatrix}, \; r_h(k-1) = \begin{bmatrix} r(k-1) \\ r(k-2) \\ \vdots \\ r(k-h) \end{bmatrix}.$$

Note that $M_{sr}^T z_{sr.h}(k-1)$ in (4.8) can be thought as an estimation of $x_s(k)$ from a deadbeat observer of S_f.

According to Subsection 3.2.1, given the value function (4.6), the optimal policy, if $x_s(k)$ were available, is

$$s(k) = -\left(W_s + D_s^T W_r D_s + \gamma B_s^T P^* B_s \right)^{-1} \left(\gamma B_s^T P^* A_s + D_s^T W_r C_s \right) x_s(k), \tag{4.9}$$

where P^* is the solution to the Riccati equation

$$P^* = \left\{ \begin{array}{l} \gamma A_s^T P^* A_s + C_s^T W_r C_s - \left(\gamma A_s^T P^* B_s + C_s^T W_r D_s \right) \\ \times \left(W_s + D_s^T W_r D_s + \gamma B_s^T P^* B_s \right)^{-1} \left(\gamma B_s^T P^* A_s + D_s^T W_r C_s \right) \end{array} \right\}, \tag{4.10}$$

and the optimal value function is $V^*(x_s(k)) = x_s^T(k) P^* x_s(k)$.

As $\boldsymbol{x}_s(k)$ is unknown, we use its deadbeat estimation (4.8) instead. Considering (4.8) and (4.9), a suboptimal policy, that is the product of an optimal state feedback gain and a deadbeat state estimation of $\boldsymbol{S}_f$, can be described by

$$s(k)=-\left(\boldsymbol{W}_s+\boldsymbol{D}_s^T\boldsymbol{W}_r\boldsymbol{D}_s+\gamma\boldsymbol{B}_s^T\boldsymbol{P}^*\boldsymbol{B}_s\right)^{-1}\left(\gamma\boldsymbol{B}_s^T\boldsymbol{P}^*\boldsymbol{A}_s+\boldsymbol{D}_s^T\boldsymbol{W}_r\boldsymbol{C}_s\right)\boldsymbol{M}_{sr}^T z_{sr.h}(k-1), \tag{4.11}$$

whose value function at k can be represented by

$$V^*\left(z_{sr.h}(k-1)\right)=z_{sr.h}^T(k-1)\,\boldsymbol{P}_z^* z_{sr.h}(k-1), \tag{4.12}$$

where $\boldsymbol{P}_z^* = \boldsymbol{M}_{sr}\boldsymbol{P}^*\boldsymbol{M}_{sr}^T$ is a constant matrix.

Remark 4.3 *Given $\boldsymbol{S}_f$ (4.7) and the value function (4.6), the optimal controller $\boldsymbol{Q}_r$ for $\boldsymbol{S}_f$ should be an $\mathcal{H}_2$ optimal controller consisting of an $\mathcal{H}_2$ optimal observer and an optimal state feedback gain. Therefore, the solution given by (4.11) is only a suboptimal one.*

To learn the suboptimal policy of $\boldsymbol{Q}_r$ with the form (4.11), we parameterize its policy as

$$s(k)=\mu\left(z_{sr.h}(k-1)\right)=\boldsymbol{\Theta}z_{sr.h}(k-1), \tag{4.13}$$

where $\boldsymbol{\Theta}$ is the parameter matrix to be optimized.

Corollary 4.1 *Given $\boldsymbol{S}_f$ (4.7) and the value function (4.6), for each stabilizing policy μ (4.13), its value function at k can be represented by*

$$V^{\mu}\left(z_{sr.h}(k-1)\right)=z_{sr.h}^T(k-1)\,\boldsymbol{P}_z^{\mu} z_{sr.h}(k-1), \tag{4.14}$$

where $\boldsymbol{P}_z^{\mu}$ is a constant matrix.

Proof Consider that both the state vector $\boldsymbol{x}(k)$ of $\boldsymbol{S}_f$ and the policy μ are represented by the past I/O data of $\boldsymbol{S}_f$ in (4.8) and (4.13), respectively. Corollary 4.1 can be derived in the same manner as Theorem 3.1. □

To learn the suboptimal control policy μ^* with the parameter matrix $\boldsymbol{\Theta}^*$ by Q-learning, the Q-function is represented, under a stabilizing policy $s(k) = \mu^i(z_{sr.h}(k-1)) = \boldsymbol{\Theta}^i z_{sr.h}(k-1)$, by

$$Q^{\mu^i}(z_{sr.h}(k-1), s(k)) = s^T(k)W_s s(k) + r^T(k)W_r r(k) + \gamma V^{\mu^i}(z_{sr.h}(k)), \tag{4.15}$$

where $s(k)$ follows a behavior policy $\mu_b \neq \mu^i$. Substituting (4.13) into (4.15), the Q-function can be rewritten by

$$\begin{aligned} Q^{\mu^i}(z_{sr.h}(k-1), s(k)) &= \begin{bmatrix} z_{sr.h}(k-1) \\ s(k) \end{bmatrix}^T \begin{bmatrix} H_{zz}^{\mu^i} & H_{zs}^{\mu^i} \\ H_{sz}^{\mu^i} & H_{ss}^{\mu^i} \end{bmatrix} \begin{bmatrix} z_{sr.h}(k-1) \\ s(k) \end{bmatrix} \\ &= \begin{bmatrix} z_{sr.h}(k-1) \\ s(k) \end{bmatrix}^T H_z^{\mu^i} \begin{bmatrix} z_{sr.h}(k-1) \\ s(k) \end{bmatrix}, \end{aligned} \tag{4.16}$$

where $H_z^{\mu^i}$ is a kernel constant matrix that relates to $P_z^{\mu^i}$. Then the recursive form of the Q-function (3.15) can be represented in the form of TD error

$$\begin{aligned} \delta(k) = &\begin{bmatrix} z_{sr.h}(k-1) \\ s(k) \end{bmatrix}^T H_z^{\mu^i} \begin{bmatrix} z_{sr.h}(k-1) \\ s(k) \end{bmatrix} \\ &- \left(s(k)^T W_s s(k) + r(k)^T W_r r(k) + \gamma \begin{bmatrix} z_{sr.h}(k) \\ \Theta^i z_{sr.h}(k) \end{bmatrix}^T H_z^{\mu^i} \begin{bmatrix} z_{sr.h}(k) \\ \Theta^i z_{sr.h}(k) \end{bmatrix} \right). \end{aligned} \tag{4.17}$$

Finally, the proposed Q-learning algorithm is given by Algorithm 4.1. When the suboptimal policy converges, it can be represented by

Algorithm 4.1 Q-learning: Robustness Optimization

1: **Initialization**: Parameterize the policy μ of $\boldsymbol{Q}_r$ with a length h of s and r data according to (4.13), and select any initial stabilizing policy μ^0 with Θ^0. Then for $i = 0, 1, \ldots$
2: **Repeat**
3: **Policy evaluation**: Identify $H_z^{\mu^i}$ using LS or SGD by minimizing the TD error (4.17) with data s, r and Θ^i.
4: **Policy improvement**: Update the policy μ^i to μ^{i+1} with $\Theta^{i+1} = -(H_{ss}^{\mu^i})^{-1} H_{sz}^{\mu^i}$.
5: **Until** the policy converges, and set it as the suboptimal policy μ^*.

$$s(k) = \Theta^* z_{sr.h}(k-1) = \Theta_s^* s_h(k-1) + \Theta_r^* r_h(k-1), \tag{4.18}$$

where $\Theta^* = \begin{bmatrix} \Theta_s^* & \Theta_r^* \end{bmatrix}$. This indicates the structure of the controller $\boldsymbol{Q}_r$. For single-input single-output (SISO), single-input multiple-output or multiple-input single-output plants, $\boldsymbol{Q_r}$ can also be described in either observable or controllable canonical form [85] for effective online implementation. Here, take an SISO plant as an example. The observable canonical form of $\boldsymbol{Q_r}$ is

$$\boldsymbol{Q_r} : \begin{cases} \boldsymbol{x_q}(k+1) = \boldsymbol{A_q x_q}(k) + \boldsymbol{b_q r}(k), \\ \boldsymbol{s}(k) = \boldsymbol{c_q x_q}(k), \end{cases} \tag{4.19}$$

where $\boldsymbol{x_q} \in \mathbb{R}^h$ is the state vector of $\boldsymbol{Q_r}$ and

$$\boldsymbol{A_q} = \begin{bmatrix} \theta_{s.1} & 1\,0 \cdots 0 \\ \theta_{s.2} & 0\,1 \cdots 0 \\ \vdots & \vdots\,\vdots\,\ddots\,\vdots \\ \theta_{s.h-1} & 0\,0 \cdots 1 \\ \theta_{s.h} & 0\,0\;0\;0 \end{bmatrix}, \boldsymbol{b_q} = \begin{bmatrix} \theta_{r.1} \\ \theta_{r.2} \\ \vdots \\ \theta_{r.h-1} \\ \theta_{r.h} \end{bmatrix}, \boldsymbol{c_q} = \begin{bmatrix} 1 \\ 0 \\ \vdots \\ 0 \\ 0 \end{bmatrix}^T,$$

and

$$\boldsymbol{\Theta_s^*} = \left[\theta_{s.1}\ \theta_{s.2} \cdots \theta_{s.h}\right], \boldsymbol{\Theta_r^*} = \left[\theta_{r.1}\ \theta_{r.2} \cdots \theta_{r.h}\right].$$

Remark 4.4 *It is favorable to choose a small length h to parameterize the policy μ in Algorithm 4.1, since the online computation during learning would be low and a low-order optimal controller could be learned. However, chances are that when the length h is too small, e.g. $h < k_s$, the policy of $\boldsymbol{Q_r}$ does not converge because Equation (4.8) does not hold. In this case, we should increment h and run the Algorithm 4.1 alternately until a converged policy is attained.*

Thus far, the purely data-driven suboptimal controller $\boldsymbol{Q_r}$ for robustness optimization is found using the proposed Q-learning algorithm. It is clear that an advantage of this optimization procedure is that it requires no model identification of $\boldsymbol{S_f}$. However, there are some issues associated with this robustness optimization strategy worth pointing out

- There is no real guidance as to the choices of weighting matrices, $\boldsymbol{W_s}$ and $\boldsymbol{W_r}$, of the performance index defined in (4.6) for robustness optimization.
- The performance loss is unknown, when $\boldsymbol{K_0}$ is designed optimally based on the performance index (3.2) for $\boldsymbol{P_0}$, and $\boldsymbol{Q_r}$ is designed optimally for $\boldsymbol{S_f}$ based on the performance index (4.6) instead of designing an optimal controller directly for $\boldsymbol{P}(\boldsymbol{S_f})$ based on (3.2).

Due to the above disadvantages, we will develop an alternative performance optimization method in the next section. This method aims to deliver an auxiliary controller that together with $\boldsymbol{K}_0$ constitutes the optimal controller for the perturbed plant $\boldsymbol{P}(\boldsymbol{S}_f)$ in terms of the prescribed performance index (3.2).

4.2.4 Simulation Results

Considering the same SISO plant P_0 described in Subsection 3.2.5, a nominal observer-based state feedback controller K_0 (2.4) is designed with the state feedback gain $\boldsymbol{F}^* = \begin{bmatrix}-1.9932 & -2.3266\end{bmatrix}$ as given in Subsection 3.2.5 and an observer gain $\boldsymbol{L} = \begin{bmatrix}0.1085 & 0.2558\end{bmatrix}^T$ based on the pole placement method.

Suppose the perturbed plant $P(S_f)$ has the following continuous-time state-space representation

$$\dot{\boldsymbol{x}}_p = \boldsymbol{A}_p\boldsymbol{x}_p + \boldsymbol{b}_p u, \quad y = \boldsymbol{c}_p\boldsymbol{x}_p,$$

$$\boldsymbol{A}_p = \begin{bmatrix} -6.6667 & 1515 & 0 & 0 \\ 0 & -81.4545 & 0 & 10 \\ -0.4 & 0 & -10 & -10 \\ 0 & 0 & 1 & 0 \end{bmatrix}, \boldsymbol{b}_p = \begin{bmatrix} 0 \\ 0 \\ 1 \\ 0 \end{bmatrix}, \boldsymbol{c}_p = \begin{bmatrix} 1 \\ 0 \\ 0 \\ 0 \end{bmatrix}^T.$$

The zero input responses of systems before and after perturbations at the same initial condition are compared and shown in Fig. 4.4. It is clear that the performance of the perturbed system degrades. The proposed IOR method is then applied to improve the system performance. The weighting matrices of the value function (4.6) are chosen to be $W_s = 0.001$ and $W_r = 10$. The length of the past I/O data, s and r, used for the parameterization of Q_r is set to 2. Fig. 4.5 shows the convergence of parameters of Q_r during online learning. The learning procedure is stopped after 10 seconds of learning. The learned Q_r (4.19) has the following system matrices

$$\boldsymbol{A}_q = \begin{bmatrix} -0.3058 & 1 \\ -0.1590 & 0 \end{bmatrix}, \boldsymbol{b}_q = \begin{bmatrix} 164.8355 \\ -109.7148 \end{bmatrix}, \boldsymbol{c}_q = \begin{bmatrix} 1 \\ 0 \end{bmatrix}^T.$$

Finally, the learned Q_r is placed back into the original closed-loop system with the structure given by Fig. 4.3. Run the closed-loop system again. Its zero input response and the one of the original closed-loop system without Q_r are compared and shown in Fig. 4.6. From Figure 4.6, it can be seen that with Q_r the perturbed system has significantly less undershoot. In addition, it can be observed from the Nyquist diagrams in Fig. 4.7 that, both the gain margin and the phase margin of the

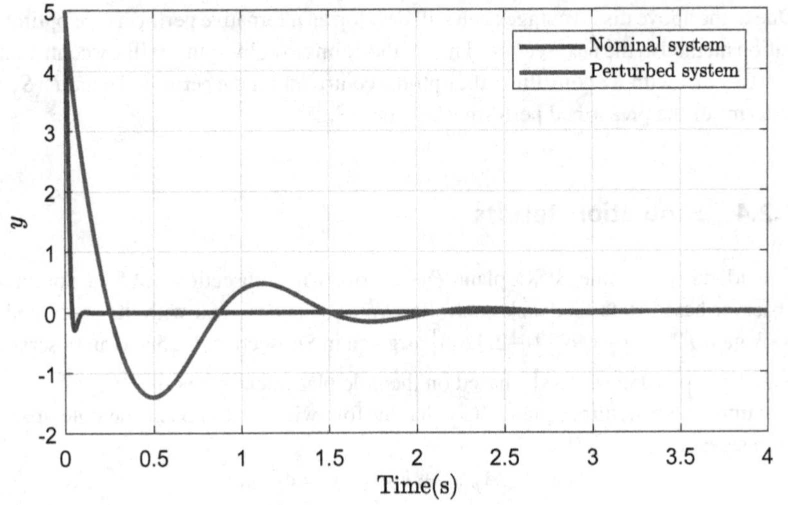

Figure 4.4 Comparison of zero input response of systems before and after perturbations

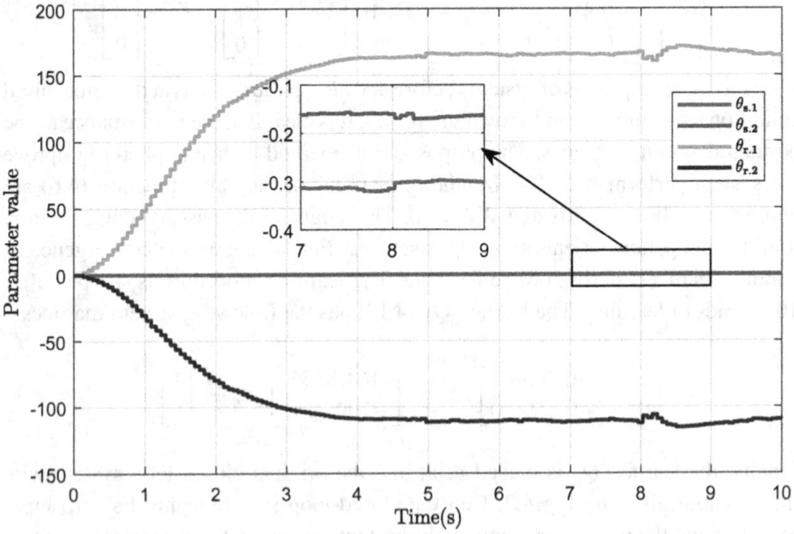

Figure 4.5 Convergence of control parameters of Q_r during online learning

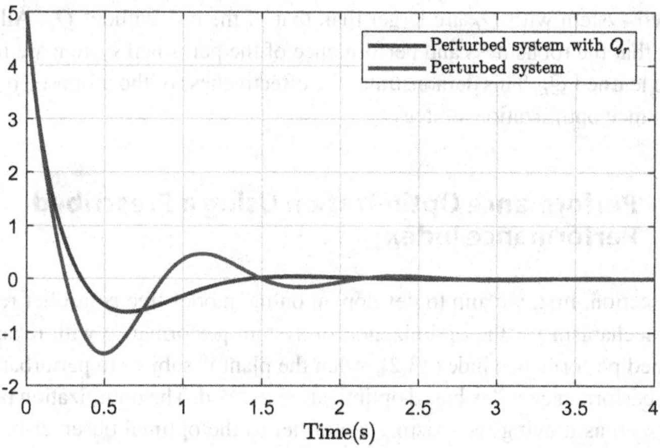

Figure 4.6 Comparison of performance of perturbed systems with and without Q_r

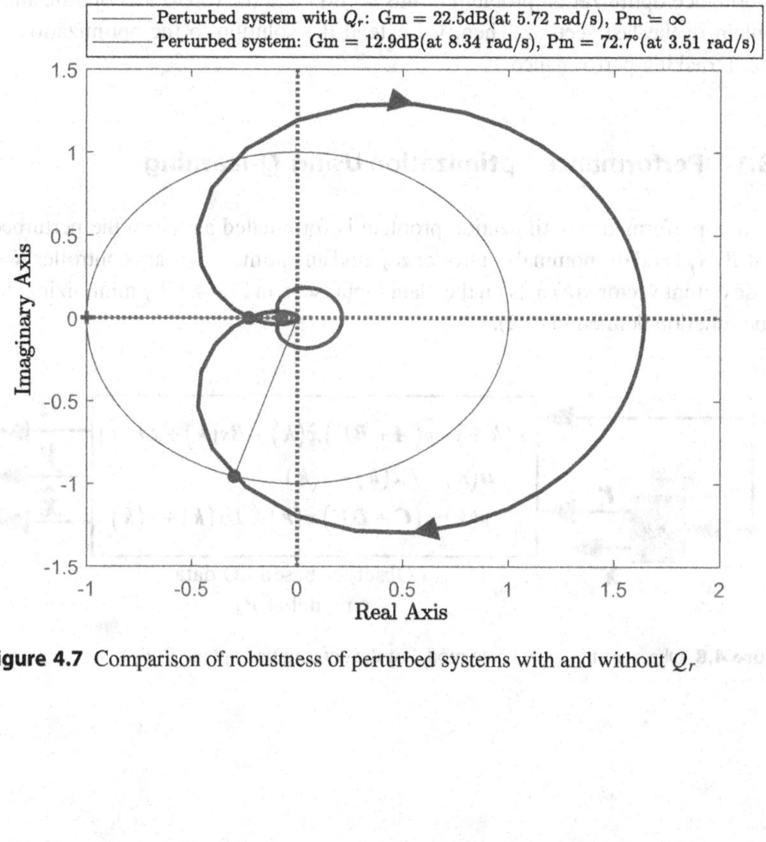

Figure 4.7 Comparison of robustness of perturbed systems with and without Q_r

perturbed system with Q_r are larger than that of the one without Q_r. All of them suggest that the robustness and performance of the perturbed system are improved with the learned Q_r. This demonstrates the effectiveness of the proposed robustness performance optimization strategy.

4.3 Performance Optimization Using a Prescribed Performance Index

In this section, first, we aim to develop an online model-free controller reconfiguration mechanism for the optimization of system performance with regard to the prescribed performance index (3.2), when the plant is subject to perturbations. We call it a performance index based optimization method. The optimization procedure can be seen as moving the existing controller to the optimal observer-based state feedback controller in the sense of $\mathcal{H}_2$ (as here the only disturbance are initial state variables) instead of LQR, seen in Fig. 4.2. This is a key difference between the performance optimization problem in this section and the robustness optimization problem in the last section. Then, we extend the solution to the optimization of system tracking performance.

4.3.1 Performance Optimization Using Q-learning

The first performance optimization problem is formulated as: given the perturbed plant $\boldsymbol{P}(\boldsymbol{S}_f)$ and the nominal controller $\boldsymbol{K}_0$, find an optimal auxiliary controller $\boldsymbol{K}_A$, whose output vector $\boldsymbol{s}(k)$ acts on the plant input, seen in Fig. 4.1, by minimizing the value function defined in (3.2).

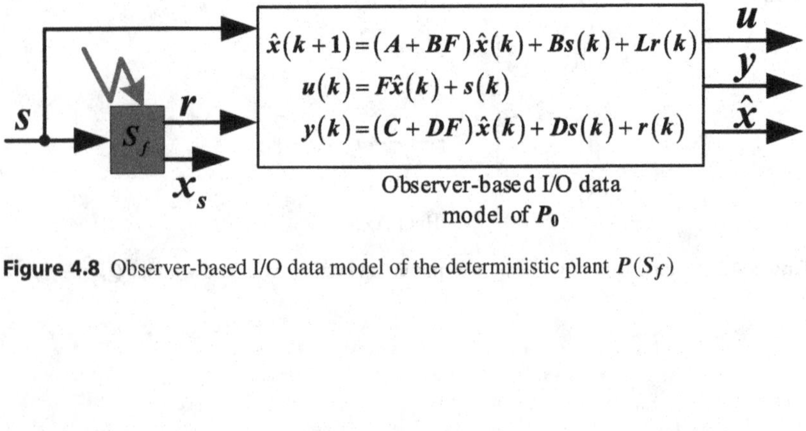

Figure 4.8 Observer-based I/O data model of the deterministic plant $\boldsymbol{P}(\boldsymbol{S}_f)$

As the perturbed plant $\boldsymbol{P}(\boldsymbol{S}_f)$ is unknown, we use its observer-based I/O model instead, which is given in (2.23). The structure is shown in Fig. 4.8. Assume that $\boldsymbol{S}_f$ has the structure described by (4.7). Then the dynamics of $\boldsymbol{P}(\boldsymbol{S}_f)$ can be represented by

$$\boldsymbol{P}(\boldsymbol{S}_f): \begin{cases} \begin{bmatrix} \hat{\boldsymbol{x}}(k+1) \\ \boldsymbol{x}_s(k+1) \end{bmatrix} = \begin{bmatrix} \boldsymbol{A}+\boldsymbol{B}\boldsymbol{F} & \boldsymbol{L}\boldsymbol{C}_s \\ \boldsymbol{0} & \boldsymbol{A}_s \end{bmatrix} \begin{bmatrix} \hat{\boldsymbol{x}}(k) \\ \boldsymbol{x}_s(k) \end{bmatrix} + \begin{bmatrix} \boldsymbol{B}+\boldsymbol{L}\boldsymbol{D}_s \\ \boldsymbol{B}_s \end{bmatrix} s(k) \\ \begin{bmatrix} \boldsymbol{u}(k) \\ \boldsymbol{y}(k) \end{bmatrix} = \begin{bmatrix} \boldsymbol{F} & \boldsymbol{0} \\ \boldsymbol{C}+\boldsymbol{D}\boldsymbol{F} & \boldsymbol{C}_s \end{bmatrix} \begin{bmatrix} \hat{\boldsymbol{x}}(k) \\ \boldsymbol{x}_s(k) \end{bmatrix} + \begin{bmatrix} \boldsymbol{I} \\ \boldsymbol{D}+\boldsymbol{D}_s \end{bmatrix} s(k) \end{cases}. \tag{4.20}$$

As the state vector $\boldsymbol{x}_s(k)$ of $\boldsymbol{S}_f$ is unknown, we represent it using the past I/O data (4.8) of $\boldsymbol{S}_f$. Combining (4.8) and (4.20), the dynamics of $\boldsymbol{u}(k)$ and $\boldsymbol{y}(k)$ is governed by $\hat{\boldsymbol{x}}(k)$ and $z_{sr.h}(k-1)$. Analogous to the results given in Subsection 4.2.3, a suboptimal policy of $\boldsymbol{K}_A$ has the form

$$s(k) = \boldsymbol{F}_a^* \hat{\boldsymbol{x}}(k) + \boldsymbol{\Theta}^* z_{sr.h}(k-1),$$

where $\boldsymbol{F}_a^*$ and $\boldsymbol{\Theta}^*$ are parameters of the suboptimal policy.

Remark 4.5 *The suboptimal policy of $\boldsymbol{K}_A$ has the property that for the perturbed plant $\boldsymbol{P}(\boldsymbol{S}_f)$, the controller consisting of the suboptimal controller $\boldsymbol{K}_A$ and the existing controller $\boldsymbol{K}_0$ is equivalent to the one consisting of a deadbeat observer and an LQR.*

The value function of the suboptimal policy has the form

$$V^*\left(\hat{\boldsymbol{x}}(k), z_{sr.h}(k-1)\right) = \begin{bmatrix} \hat{\boldsymbol{x}}(k) \\ z_{sr.h}(k-1) \end{bmatrix}^T \boldsymbol{P}_p^* \begin{bmatrix} \hat{\boldsymbol{x}}(k) \\ z_{sr.h}(k-1) \end{bmatrix}, \tag{4.21}$$

where $\boldsymbol{P}_p^*$ is a constant matrix. Accordingly, the policy of $\boldsymbol{K}_A$ can be parameterized by

$$s(k) = \mu\left(\hat{\boldsymbol{x}}(k), z_{sr.h}(k-1)\right) = \boldsymbol{F}_a \hat{\boldsymbol{x}}(k) + \boldsymbol{\Theta} z_{sr.h}(k-1), \tag{4.22}$$

where $\boldsymbol{F}_a$ and $\boldsymbol{\Theta}$ are parameters to be optimized.

Corollary 4.2 *Given $\boldsymbol{P}(\boldsymbol{S}_f)$ (4.20), the nominal controller $\boldsymbol{K}_0$ (2.4) and the value function V^μ (3.2), for each stabilizing policy μ (4.22), its value function at k has the form*

$$V^{\mu}\left(\hat{x}(k), z_{sr.h}(k-1)\right) = \begin{bmatrix} \hat{x}(k) \\ z_{sr.h}(k-1) \end{bmatrix}^T P_p^{\mu} \begin{bmatrix} \hat{x}(k) \\ z_{sr.h}(k-1) \end{bmatrix}, \tag{4.23}$$

where P_p^{μ} is a constant matrix.

Proof Corollary 4.2 can derived in the same manner as Theorem 3.1. □

To learn the suboptimal control policy μ^* with parameters F_a^* and Θ^* by Q-learning, we represent the Q-function, under a stabilizing policy $s(k) = \mu^i(\hat{x}(k), z_{sr.h}(k-1)) = F_a^i \hat{x}(k) + \Theta^i z_{sr.h}(k-1)$, by

$$\begin{aligned} &Q^{\mu^i}(\hat{x}(k), z_{sr.h}(k-1), s(k)) \\ &= u(k)^T R u(k) + y(k)^T Q y(k) + \gamma V^{\mu^i}\left(\hat{x}(k+1), z_{sr.h}(k)\right), \end{aligned} \tag{4.24}$$

where $u(k) = F\hat{x}(k) + s(k)$ and $s(k)$ follows a behavior policy $\mu_b \neq \mu^i$.

Considering (4.8) and (4.20), the Q-function can be rewritten by

$$\begin{aligned} Q^{\mu^i}(\hat{x}(k), z_{sr.h}(k-1), s(k)) &= \begin{bmatrix} \hat{x}(k) \\ z_{sr.h}(k-1) \\ s(k) \end{bmatrix}^T \begin{bmatrix} H_{xx}^{\mu^i} & H_{xz}^{\mu^i} & H_{xs}^{\mu^i} \\ H_{zx}^{\mu^i} & H_{zz}^{\mu^i} & H_{zs}^{\mu^i} \\ H_{sx}^{\mu^i} & H_{sz}^{\mu^i} & H_{ss}^{\mu^i} \end{bmatrix} \begin{bmatrix} \hat{x}(k) \\ z_{sr.h}(k-1) \\ s(k) \end{bmatrix} \\ &= \begin{bmatrix} \hat{x}(k) \\ z_{sr.h}(k-1) \\ s(k) \end{bmatrix}^T H_p^{\mu^i} \begin{bmatrix} \hat{x}(k) \\ z_{sr.h}(k-1) \\ s(k) \end{bmatrix}, \end{aligned} \tag{4.25}$$

where $H_p^{\mu^i}$ is a constant matrix. Then the TD error for identifying $H_p^{\mu^i}$ is described by

$$\begin{aligned} \delta(k) = &\begin{bmatrix} \hat{x}(k) \\ z_{sr.h}(k-1) \\ s(k) \end{bmatrix}^T H_p^{\mu^i} \begin{bmatrix} \hat{x}(k) \\ z_{sr.h}(k-1) \\ s(k) \end{bmatrix} - u(k)^T R u(k) - y(k)^T Q y(k) \\ &- \gamma \begin{bmatrix} \hat{x}(k+1) \\ z_{sr.h}(k) \\ F_a^i \hat{x}(k+1) + \Theta^i z_{sr.h}(k) \end{bmatrix}^T H_p^{\mu^i} \begin{bmatrix} \hat{x}(k+1) \\ z_{sr.h}(k) \\ F_a^i \hat{x}(k+1) + \Theta^i z_{sr.h}(k) \end{bmatrix}. \end{aligned} \tag{4.26}$$

Finally, the implementation of the proposed performance optimization strategy using Q-learning is elaborated in Algorithm 4.2.

To sum up, Table 4.1 shows a comparison of the two developed performance optimization methods, the IOR method and the performance index based optimization method.

Algorithm 4.2 Q-learning: Performance Optimization Based on (3.2)

1: **Initialization**: Parameterize the policy μ with a length h of $\boldsymbol{s}$ and $\boldsymbol{r}$ data according to (4.13), and select any initial stabilizing policy μ^0 with $\boldsymbol{F}_a^0$ and $\boldsymbol{\Theta}^0$. Then for $i = 0, 1, \ldots$
2: **Repeat**
3: **Policy evaluation**: Identify $\boldsymbol{H}_p^{\mu^i}$ using LS or SGD by minimizing the TD error (4.26) with data $\boldsymbol{s}, \boldsymbol{r}, \hat{\boldsymbol{x}}, \boldsymbol{u}, \boldsymbol{y}$, and $\boldsymbol{F}_a^i$ and $\boldsymbol{\Theta}^i$.
4: **Policy improvement**: Update the policy μ^i to μ^{i+1} with $\boldsymbol{F}_a^{i+1} = -(\boldsymbol{H}_{ss}^{\mu^i})^{-1}\boldsymbol{H}_{sx}^{\mu^i}$ and $\boldsymbol{\Theta}^{i+1} = -(\boldsymbol{H}_{ss}^{\mu^i})^{-1}\boldsymbol{H}_{sz}^{\mu^i}$.
5: **Until** the policy converges, and set it as the suboptimal policy μ^*.

Table 4.1 Comparison of IOR and performance index based optimization methods

Methods	Performance objective	Robustness objective	Disadvantages
IOR	LQR optimal ($\boldsymbol{P}_0$ and optimal state feedback gain)	LQR robustness	• Uncertain weighting matrices of (4.6) • Unclear optimality
Performance index based	$\mathcal{H}_2$ optimal ($\boldsymbol{P}(\boldsymbol{S}_f)$ and optimal feedback controller)	Not assigned	• Slow learning • No robustness guarantee

4.3.2 An Extension to Tracking Performance Optimization

In this subsection, we further extend the performance index based optimization method in the previous subsection to solve a tracking problem.

Given the nominal plant model $\boldsymbol{P}_0$ (2.1) and the following reference input model

$$\begin{cases} \boldsymbol{x}_t(k+1) = \boldsymbol{A}_t\boldsymbol{x}_t(k), \quad \boldsymbol{x}_t(0) \\ \boldsymbol{y}_{ref}(k) = \boldsymbol{C}_t\boldsymbol{x}_t(k), \end{cases} \tag{4.27}$$

where $\boldsymbol{x}_t(k) \in \mathbb{R}^{k_t}$ and $\boldsymbol{y}_{ref}(k) \in \mathbb{R}^m$ are the state and the reference input vectors. The initial state is $\boldsymbol{x}_t(0)$. $\boldsymbol{A}_t$ and $\boldsymbol{C}_t$ are known parameter matrices. Assume that $(\boldsymbol{C}_t, \boldsymbol{A}_t)$ is observable and both $\boldsymbol{x}_t(k)$ and $\boldsymbol{y}_{ref}(k)$ are measurable. We define a performance index by the following value function

$$V(k) = \lim_{N\to\infty} \sum_{i=k}^{N} \gamma^{i-k}\left(\boldsymbol{u}^T(i)\boldsymbol{W}_u\boldsymbol{u}(i) + \tilde{\boldsymbol{y}}^T(i)\boldsymbol{W}_{\tilde{y}}\tilde{\boldsymbol{y}}(i)\right),$$
$$\tilde{\boldsymbol{y}}(i) = \boldsymbol{y}(i) - \boldsymbol{y}_{ref}(i), \tag{4.28}$$

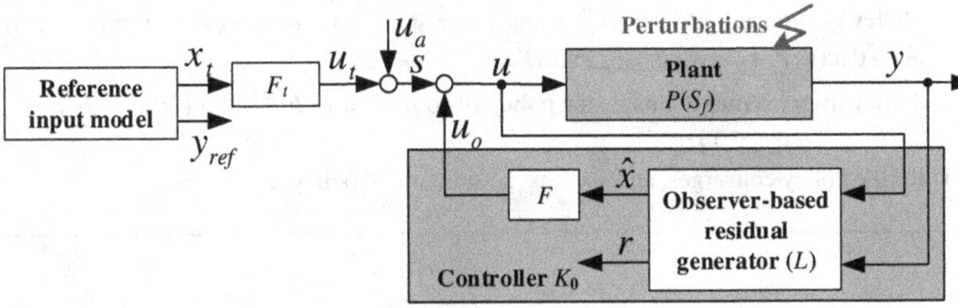

Figure 4.9 Tracking control structure

where $\boldsymbol{W}_u > \boldsymbol{0}$ and $\boldsymbol{W}_{\tilde{y}} \geq \boldsymbol{0}$ are weighting matrices for the control energy and the tracking error, respectively.

Suppose that initially a nominal feedback controller $\boldsymbol{K}_0$ (2.4) and a feedforward controller with a feedforward gain $\boldsymbol{F}_t$ are designed for $\boldsymbol{P}_0$ to guarantee stability and good tracking performance. Assume that perturbations happen in the plant degrade the tracking performance and have no influence on stability. We represent the perturbed plant as $\boldsymbol{P}(\boldsymbol{S}_f)$. Our objective is to design a strategy $\boldsymbol{u}_a$ acting on the plant input, so as to recover the good tracking performance based on the value function (4.28). The control structure is shown in Fig. 4.9.

The tracking performance optimization problem can be easily solved in a similar procedure as given in the previous subsection by considering the reference input model (4.27) as a part of the plant. Then, the suboptimal control strategy $\boldsymbol{u}_a$ has

$$\boldsymbol{u}_a(k) = \boldsymbol{F}_{ta}^*\boldsymbol{x}_t(k) + \boldsymbol{F}_a^*\hat{\boldsymbol{x}}(k) + \boldsymbol{\Theta}^* z_{sr.h}(k-1). \tag{4.29}$$

where $\boldsymbol{F}_{ta}^*$, $\boldsymbol{F}_a^*$, and $\boldsymbol{\Theta}^*$ are control parameters. The associated value function of the suboptimal control strategy $\boldsymbol{u}_a$ has the form

$$V^* \left(\boldsymbol{x}_t(k), \hat{\boldsymbol{x}}(k), z_{sr.h}(k-1)\right) = \begin{bmatrix} \boldsymbol{x}_t(k) \\ \hat{\boldsymbol{x}}(k) \\ z_{sr.h}(k-1) \end{bmatrix}^T \boldsymbol{P}_t^* \begin{bmatrix} \boldsymbol{x}_t(k) \\ \hat{\boldsymbol{x}}(k) \\ z_{sr.h}(k-1) \end{bmatrix}, \quad (4.30)$$

where $\boldsymbol{P}_t^*$ is a constant matrix. Accordingly, the policy for tracking performance optimization can be parameterized by

$$\boldsymbol{u}_a(k) = \mu\left(\boldsymbol{x}_t(k), \hat{\boldsymbol{x}}(k), z_{sr.h}(k-1)\right) = \boldsymbol{F}_{ta}\boldsymbol{x}_t(k) + \boldsymbol{F}_a\hat{\boldsymbol{x}}(k) + \boldsymbol{\Theta} z_{sr.h}(k-1), \quad (4.31)$$

where $\boldsymbol{F}_{ta}$, $\boldsymbol{F}_a$ and $\boldsymbol{\Theta}$ are parameters to be optimized.

To learn the suboptimal control policy μ^* with optimal parameters $\boldsymbol{F}_{ta}^*$, $\boldsymbol{F}_a^*$ and $\boldsymbol{\Theta}^*$ by Q-learning, we represent Q-function, under a stabilizing policy $\boldsymbol{u}_a(k) = \mu^i(\boldsymbol{x}_t(k), \hat{\boldsymbol{x}}(k), z_{sr.h}(k-1)) = \boldsymbol{F}_{ta}^i\boldsymbol{x}_t(k) + \boldsymbol{F}_a^i\hat{\boldsymbol{x}}(k) + \boldsymbol{\Theta}^i z_{sr.h}(k-1)$, by

$$\begin{aligned} Q^{\mu^i}(\boldsymbol{x}_t(k), \hat{\boldsymbol{x}}(k), z_{sr.h}(k-1), \boldsymbol{u}_a(k)) =& \boldsymbol{u}(k)^T \boldsymbol{W}_u \boldsymbol{u}(k) + \tilde{\boldsymbol{y}}(k)^T \boldsymbol{W}_{\tilde{y}} \tilde{\boldsymbol{y}}(k) \\ &+ \gamma V^{\mu^i}\left(\boldsymbol{x}_t(k+1), \hat{\boldsymbol{x}}(k+1), z_{sr.h}(k)\right), \end{aligned} \quad (4.32)$$

where $\boldsymbol{u}(k) = \boldsymbol{F}_t\boldsymbol{x}_t(k) + \boldsymbol{F}\hat{\boldsymbol{x}}(k) + \boldsymbol{u}_a(k)$ and $\boldsymbol{u}_a(k)$ follows a behavior policy $\mu_b \neq \mu^i$. Considering (4.8), (4.20) and (4.27), the Q-function can be rewritten by

$$\begin{aligned} &Q^{\mu^i}(\boldsymbol{x}_t(k), \hat{\boldsymbol{x}}(k), z_{sr.h}(k-1), \boldsymbol{u}_a(k)) \\ &= \begin{bmatrix} \boldsymbol{x}_t(k) \\ \hat{\boldsymbol{x}}(k) \\ z_{sr.h}(k-1) \\ \boldsymbol{u}_a(k) \end{bmatrix}^T \begin{bmatrix} \boldsymbol{H}_{tt}^{\mu^i} & \boldsymbol{H}_{tx}^{\mu^i} & \boldsymbol{H}_{tz}^{\mu^i} & \boldsymbol{H}_{tu}^{\mu^i} \\ \boldsymbol{H}_{xt}^{\mu^i} & \boldsymbol{H}_{xx}^{\mu^i} & \boldsymbol{H}_{xz}^{\mu^i} & \boldsymbol{H}_{xu}^{\mu^i} \\ \boldsymbol{H}_{zt}^{\mu^i} & \boldsymbol{H}_{zx}^{\mu^i} & \boldsymbol{H}_{zz}^{\mu^i} & \boldsymbol{H}_{zu}^{\mu^i} \\ \boldsymbol{H}_{ut}^{\mu^i} & \boldsymbol{H}_{ux}^{\mu^i} & \boldsymbol{H}_{uz}^{\mu^i} & \boldsymbol{H}_{uu}^{\mu^i} \end{bmatrix} \begin{bmatrix} \boldsymbol{x}_t(k) \\ \hat{\boldsymbol{x}}(k) \\ z_{sr.h}(k-1) \\ \boldsymbol{u}_a(k) \end{bmatrix} \\ &= \begin{bmatrix} \boldsymbol{x}_t(k) \\ \hat{\boldsymbol{x}}(k) \\ z_{sr.h}(k-1) \\ \boldsymbol{u}_a(k) \end{bmatrix}^T \boldsymbol{H}_t^{\mu^i} \begin{bmatrix} \boldsymbol{x}_t(k) \\ \hat{\boldsymbol{x}}(k) \\ z_{sr.h}(k-1) \\ \boldsymbol{u}_a(k) \end{bmatrix}, \end{aligned}$$

where $\boldsymbol{H}_t^{\mu^i}$ is a constant matrix. Then the TD error for identifying $\boldsymbol{H}_t^{\mu^i}$ is described by

$$\delta(k)=\begin{bmatrix} x_t(k) \\ \hat{x}(k) \\ z_{sr.h}(k-1) \\ u_a(k) \end{bmatrix}^T H_t^{\mu^i} \begin{bmatrix} x_t(k) \\ \hat{x}(k) \\ z_{sr.h}(k-1) \\ u_a(k) \end{bmatrix} - u(k)^T W_u u(k) - \tilde{y}(k)^T W_{\tilde{y}} \tilde{y}(k)$$

$$-\gamma \begin{bmatrix} x_t(k+1) \\ \hat{x}(k+1) \\ z_{sr.h}(k) \\ u_a^i(k+1) \end{bmatrix}^T H_t^{\mu^i} \begin{bmatrix} x_t(k+1) \\ \hat{x}(k+1) \\ z_{sr.h}(k) \\ u_a^i(k+1) \end{bmatrix}, \tag{4.33}$$

$$u_a^i(k+1) = F_{ta}^i x_t(k+1) + F_a^i \hat{x}(k+1) + \Theta^i z_{sr.h}(k).$$

Finally, the proposed Q-learning strategy for tracking performance optimization is elaborated in Algorithm 4.3.

Algorithm 4.3 Q-learning: Tracking Performance Optimization Based on (4.28)

1: **Initialization**: Parameterize the policy μ with a length h of s and r data according to (4.13), and select any initial stabilizing policy μ^0 with F_{ta}^0, F_a^0 and Θ^0. Then for $i = 0, 1, \ldots$

2: **Repeat**

3: **Policy evaluation**: Identify $H_t^{\mu^i}$ using LS or SGD by minimizing the TD error (4.33) with data $s, r, \hat{x}, x_t, u, u_a, y, y_{ref}$ and F_{ta}^i, F_a^i and Θ^i.

4: **Policy improvement**: Update the policy μ^i to μ^{i+1} with $F_{ta}^{i+1} = -(H_{uu}^{\mu^i})^{-1} H_{ut}^{\mu^i}$, and $F_a^{i+1} = -(H_{uu}^{\mu^i})^{-1} H_{ux}^{\mu^i}$ and $\Theta^{i+1} = -(H_{uu}^{\mu^i})^{-1} H_{uz}^{\mu^i}$.

5: **Until** the policy converges, and set it as the suboptimal policy μ^*.

4.3.3 Simulation Results

In this subsection, a simulation study on an inverted pendulum system is used to show the feasibility of the proposed performance index based optimization approach.

The laboratory inverted pendulum system LIP100 is sketched in Fig. 4.10. It consists of a cart (position 1) that moves along a metal track (position 2), and a rod (position 3) with a cylindrical weight (position 4) that can pivot on the cart. The cart is driven by a DC motor, whose torque is proportional to its voltage input. Detailed parameters of the system can be found in [13]. Based on the system, a linearized model around the upright position of the rod is used in this simulation study, which is described by

$$\dot{x} = Ax + bu, \quad y = Cx, \tag{4.34}$$

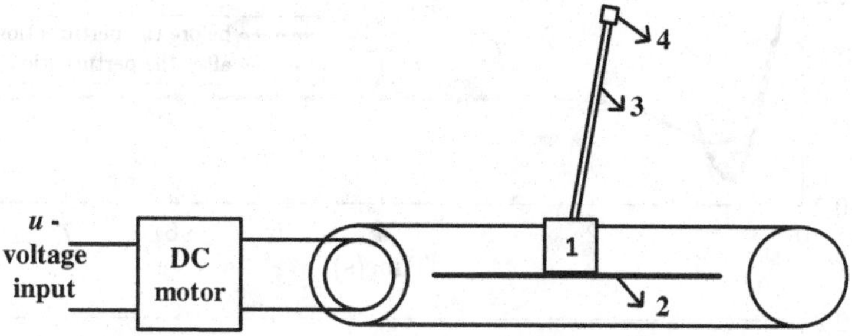

Figure 4.10 A schematic sketch of an inverted pendulum system

$$A = \begin{bmatrix} 0 & 0 & 1 & 0 \\ 0 & 0 & 0 & 1 \\ 0 & -0.8800 & -1.9146 & 0.0056 \\ 0 & 21.4711 & 3.8452 & -0.1362 \end{bmatrix}, \boldsymbol{b} = \begin{bmatrix} 0 \\ 0 \\ 0.6948 \\ -1.3955 \end{bmatrix}, \boldsymbol{C} = \begin{bmatrix} 1 & 0 & 0 & 0 \\ 0 & 1 & 0 & 0 \end{bmatrix},$$

where the state $\boldsymbol{x}$ consists of the cart position, the rod angle, the cart speed, and the rod speed. u is the voltage input, and $\boldsymbol{y}$ is the output vector consisting of the cart position and the rod angle. The performance index is defined in (3.2) with the weighting matrices

$$R = 0.1, \quad \boldsymbol{Q} = \begin{bmatrix} 40 & 0 \\ 0 & 40 \end{bmatrix},$$

and the discount factor $\gamma = 0.9999$ and the sampling time is 0.02 s.

Initially, an observer-based state feedback nominal controller $\boldsymbol{K}_0$ is designed with the optimal state feedback gain $\boldsymbol{F}^*$ based on the defined cost function and a deadbeat observer gain $\boldsymbol{L}$ to guarantee stability and good performance. They are

$$\boldsymbol{F}^* = \begin{bmatrix} 19.9991 & 92.2050 & 20.6656 & 19.692 \end{bmatrix},$$

$$\boldsymbol{L} = \begin{bmatrix} 1.9624 & 0.0754 & 47.2060 & 5.6122 \\ -2.3674 \times 10^{-4} & 2.0058 & -0.0235 & 50.5815 \end{bmatrix}^T.$$

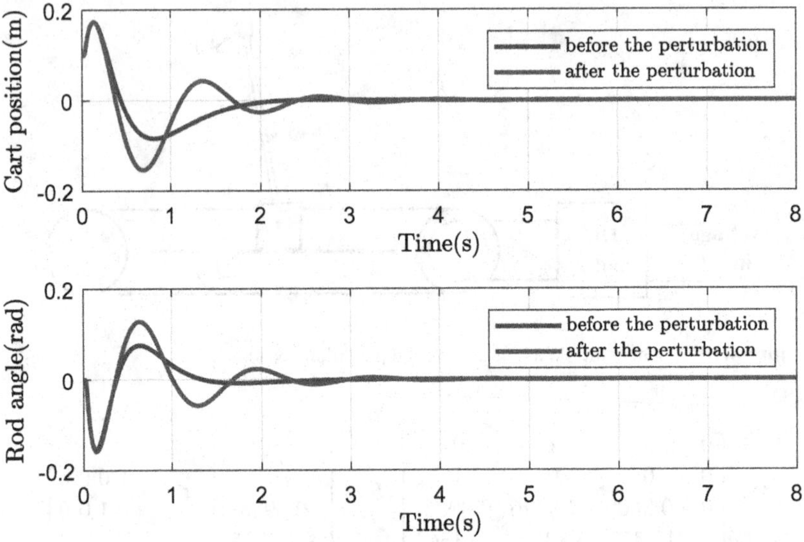

Figure 4.11 Comparison of performance of inverted pendulum systems before and after the perturbation

Then, a change of the mass of the cylindrical weight from 3.529 kg to 4 kg perturbs the existing inverted pendulum system. The perturbed plant $\boldsymbol{P}(\boldsymbol{S}_f)$ has the same state-space form as (4.34) with the system matrix $\boldsymbol{A}$ changed to

$$\boldsymbol{A}=\begin{bmatrix} 0 & 0 & 1 & 0 \\ 0 & 0 & 0 & 1 \\ 0 & -0.8800 & -1.9146 & 0.0056 \\ 0 & 24.3367 & 3.8452 & -0.1544 \end{bmatrix}.$$

System matrices $\boldsymbol{b}$ and $\boldsymbol{C}$ remain unchanged. Fig. 4.11 shows the comparison of system performance before and after the perturbation. It is clear that system performance degrades when the perturbation enters. To improve the performance, the performance index based method is applied to learn the optimal auxiliary controller $\boldsymbol{K}_A$. We parameterize its policy (4.22) using s and r data with a length $h = 2$. Fig. 4.12 shows the convergence of $\boldsymbol{\Theta}$ and $\boldsymbol{F}_a$ during online learning. The learned parameters of $\boldsymbol{K}_A$ are

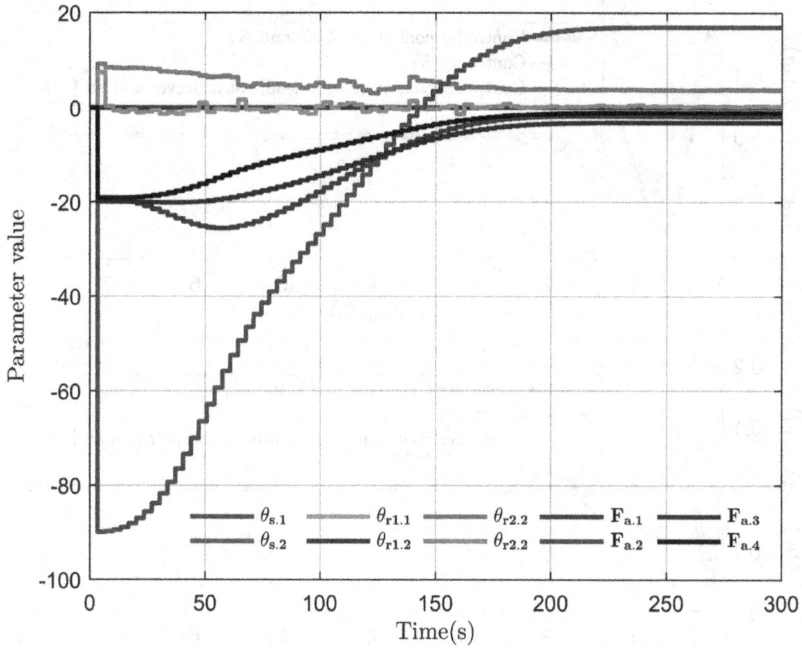

Figure 4.12 Convergence of $\boldsymbol{\Theta}$ and $\boldsymbol{F_a}$ during online learning

$$\boldsymbol{\Theta} = \left[-0.002491\ -0.0002524\ 0.0002814\ -0.001391\ 3.741\ 0.2564\right],$$
$$\boldsymbol{F_a} = \left[-1.934\ 16.95\ -3.188\ -1.144\right].$$

Finally, Fig. 4.13 shows the comparison of performance of systems consisting of $\boldsymbol{P}(\boldsymbol{S_f})$ and different controllers. In particular, the green curve in Fig. 4.13 shows the response of the system consisting of $\boldsymbol{P}(\boldsymbol{S_f})$ and the theoretically computed suboptimal controller consisting of a deadbeat observer for $\boldsymbol{P}(\boldsymbol{S_f})$ and an LQR based on the defined performance index. It is clear from Fig. 4.13 that the auxiliary controller $\boldsymbol{K_A}$ improves the system performance. In addition, the controller consisting of $\boldsymbol{K_0}$ and $\boldsymbol{K_A}$ is equivalent to the one consisting of a deadbeat observer and an LQR. This is the property of the learned controller, as pointed out in Remark 4.5, when the proposed Q-learning method is applied.

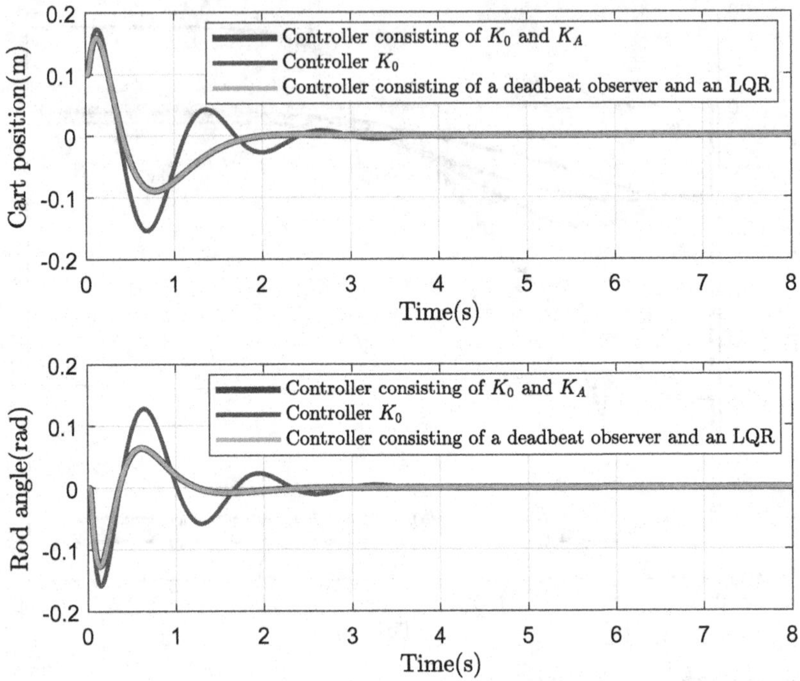

Figure 4.13 Comparison of zero input response of systems consisting of $\boldsymbol{P}(\boldsymbol{S}_f)$ and different controllers

4.4 Concluding Remarks

In this chapter, we have dealt with performance degradation problems of deterministic systems from two different viewpoints: robustness optimization and performance optimization using a prescribed performance index.

As to robustness optimization, we have proposed a new IOR method, which provides an alternative approach to many existing methods, such as LTR. The key idea of IOR is to design an auxiliary controller $\boldsymbol{Q}_r$ to recover gracefully the I/O responses of the system consisting of the faulty plant $\boldsymbol{P}(\boldsymbol{S}_f)$ and the new controller $\boldsymbol{K}(\boldsymbol{Q}_r)$ to those of the system consisting of the nominal plant $\boldsymbol{P}_0$ and an LQR. As the latter system has good performance and robustness guarantees, this IOR strategy promises good optimization results.

The proposed second optimization approach aims to find an auxiliary controller to optimize the system performance by minimizing the value function, based on which the nominal controller $\boldsymbol{K}_0$ is built. Thus, the optimal auxiliary controller has the property that it together with the nominal controller $\boldsymbol{K}_0$ constitutes the optimal controller for the perturbed plant $\boldsymbol{P}(\boldsymbol{S}_f)$ in terms of the prescribed performance index. We have also extended this performance optimization approach to solve a tracking performance degradation problem.

We would emphasize that in all cases, we have found suboptimal controllers by Q-learning in a data-driven manner and there has been no involvement of system identification.

5 NAC Aided Performance Optimization of Stochastic Systems

In the last chapter, we have investigated performance degradation problems of deterministic systems. A new IOR method and a performance index based method have been proposed for system performance optimization and implemented by Q-learning. This shows successful applications of RL methods to performance optimization of feedback control systems.

In this chapter, we take a step further and study performance degradation problems in a stochastic environment. To solve them, we again adopt the IOR and performance index based methods. But for implementation we develop new data-driven approaches based on the on-policy NAC policy gradient method proposed in Chapter 3. The reason is that Q-learning, which relies heavily on the accurate identification of the accurately parameterized value function, tends to fail in a stochastic environment.

In addition, so far, we have restricted ourselves to studying performance optimization methods for plants with observer-based state feedback controllers, where a residual vector is available. This is, however, not the case in many industrial applications, such as in occasions where PID controllers are applied. Motivated by this, we will also develop an approach for performance optimization of plants with general output feedback controllers.

This chapter is then organized as follows. First, we apply the IOR method to system robustness optimization, and study noise characteristics so that the NAC method can be applied to data-driven implementation. We also discuss conditions for closed-loop stability during online learning. Second, we apply the performance index based method to system performance optimization, and study a data-efficient

C. Hua, *Reinforcement Learning Aided Performance Optimization of Feedback Control Systems*, https://doi.org/10.1007/978-3-658-33034-7_5

learning approach using both plant nominal model and real-time data for data-driven implementation. Finally, we generalize the performance index based method to performance optimization of plants with general output feedback controllers and demonstrate its effectiveness by a benchmark study on a brushless direct current (BLDC) motor test rig.

5.1 Problem Formulation

We consider the stochastic plant $\boldsymbol{P}$ described in (3.30). Assume that

- An initial nominal observer-based state feedback controller $\boldsymbol{K_0}$ (2.4) is designed for $\boldsymbol{P}$ guaranteeing stability and good performance based on the value function defined in (3.33).
- Perturbations that happen in the plant degrade the performance and have no effect on system stability. According to Theorem 2.5, we represent the perturbed stochastic plant by $\boldsymbol{P}(\boldsymbol{S_f})$ with $\boldsymbol{S_f} \in \mathcal{RH}_\infty$ denoting the perturbations.

The initial control structure is shown in Fig. 5.1 where $s = \mathbf{0}$. The objective is to develop online controller reconfiguration mechanisms to optimize the system performance satisfying the three conditions given in Section 4.1.

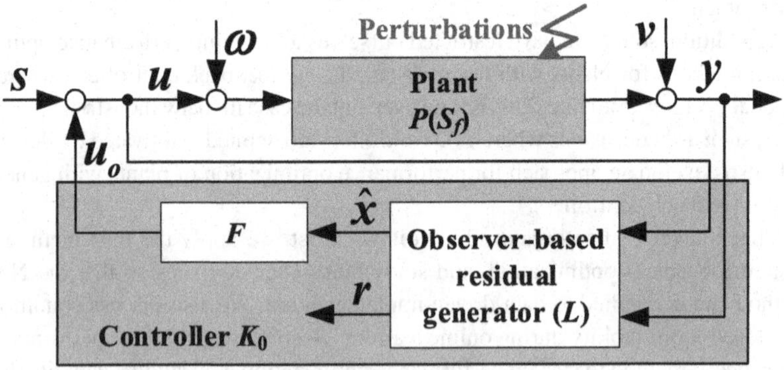

Figure 5.1 The system configuration after perturbations

5.2 Robustness Optimization

It is evident that the proposed IOR method in Chapter 4 can be used to improve the system robustness. An auxiliary controller $\boldsymbol{Q}_r$ that feeds back the residual vector $\boldsymbol{r}$ to the control input $\boldsymbol{s}$ is the design parameter. The optimization problem can then be formulated as finding $\boldsymbol{Q}_r$ that minimizes the following value function

$$V(k) = \lim_{N\to\infty} \mathbb{E}\left(\sum_{i=k}^{N} \gamma^{i-k}\left(\boldsymbol{s}^T(i)\boldsymbol{W}_s\boldsymbol{s}(i) + \boldsymbol{r}^T(i)\boldsymbol{W}_r\boldsymbol{r}(i)\right)\right). \tag{5.1}$$

5.2.1 Noise Characteristics

The optimization of $\boldsymbol{Q}_r$ based on the value function (5.1) where $\boldsymbol{s}$ and $\boldsymbol{r}$ are input and output vectors of the unknown plant $\boldsymbol{S}_f$ is similar to the optimization of an LQG controller in Subsection 3.3.2, which is solved by an on-policy NAC method. Only the noise characteristics are still unclear. Therefore, the main task of this subsection is to check the noise characteristics of the closed-loop pair $(\boldsymbol{S}_f, \boldsymbol{Q}_r)$ to decide in which condition the NAC method proposed in Chapter 3 can be applied for the optimization of $\boldsymbol{Q}_r$.

It is known from (2.16) that

$$\begin{bmatrix} \boldsymbol{s} \\ \boldsymbol{r} \end{bmatrix} = \begin{bmatrix} \boldsymbol{X} & \boldsymbol{Y} \\ -\hat{\boldsymbol{N}} & \hat{\boldsymbol{M}} \end{bmatrix} \begin{bmatrix} \boldsymbol{u} \\ \boldsymbol{y} \end{bmatrix}, \tag{5.2}$$

then we have

$$\boldsymbol{r} - \boldsymbol{S}_f\boldsymbol{s} = \left(\hat{\boldsymbol{M}} - \boldsymbol{S}_f\boldsymbol{Y}\right)\boldsymbol{y} - \left(\hat{\boldsymbol{N}} + \boldsymbol{S}_f\boldsymbol{X}\right)\boldsymbol{u}. \tag{5.3}$$

The I/O relationship of $\boldsymbol{P}(\boldsymbol{S}_f)$ is

$$\boldsymbol{y} = \boldsymbol{P}\left(\boldsymbol{S}_f\right)\left(\boldsymbol{u} + \boldsymbol{\omega}\right) + \boldsymbol{v}. \tag{5.4}$$

Substituting (5.4) into (5.3) and considering (2.21), the relationship between $\boldsymbol{s}(k)$ and $\boldsymbol{r}(k)$ can be established by

$$\boldsymbol{r} = \boldsymbol{S}_f\left(\boldsymbol{s} + \boldsymbol{n}_\omega\right) + \boldsymbol{n}_v, \tag{5.5}$$

where $\boldsymbol{n}_\omega(k)$ and $\boldsymbol{n}_v(k)$ have

$$\begin{bmatrix} \boldsymbol{n}_\omega \\ \boldsymbol{n}_v \end{bmatrix} = \begin{bmatrix} X & -Y \\ \hat{N} & \hat{M} \end{bmatrix} \begin{bmatrix} \boldsymbol{\omega} \\ \boldsymbol{v} \end{bmatrix}. \tag{5.6}$$

This shows that the effect of $\boldsymbol{\omega}$ and $\boldsymbol{v}$ acting on $\boldsymbol{u}$ and $\boldsymbol{y}$ is equivalent to that of $\boldsymbol{n}_\omega$ and $\boldsymbol{n}_v$ acting on $\boldsymbol{s}$ and $\boldsymbol{r}$. Finally, considering Equation (2.21) and the above noise characteristics, the LCF and RCF of the stochastic plant $\boldsymbol{P}(\boldsymbol{S}_f)$ are shown in Fig. 5.2.

In accord with (5.5), $\boldsymbol{S}_f$ can be represented in a state-space form by

$$\boldsymbol{S}_f : \begin{cases} \boldsymbol{x}_s(k+1) = \boldsymbol{A}_s \boldsymbol{x}_s(k) + \boldsymbol{B}_s\big(\boldsymbol{s}(k) + \boldsymbol{n}_\omega(k)\big), \\ \boldsymbol{r}(k) = \boldsymbol{C}_s \boldsymbol{x}_s(k) + \boldsymbol{D}_s\big(\boldsymbol{s}(k) + \boldsymbol{n}_\omega(k)\big) + \boldsymbol{n}_v(k), \end{cases} \tag{5.7}$$

where $\boldsymbol{x}_s(k) \in \mathbb{R}^{k_s}$ represents the state vector. $\boldsymbol{A}_s$, $\boldsymbol{B}_s$, $\boldsymbol{C}_s$ and $\boldsymbol{D}_s$ are unknown constant matrices with proper dimensions.

Note that the application of the NAC method proposed in Chapter 3 to the optimization of $\boldsymbol{Q}_r$ based on the value function (5.1) requires that $\boldsymbol{n}_\omega$ and $\boldsymbol{n}_v$ should be Gaussian white.

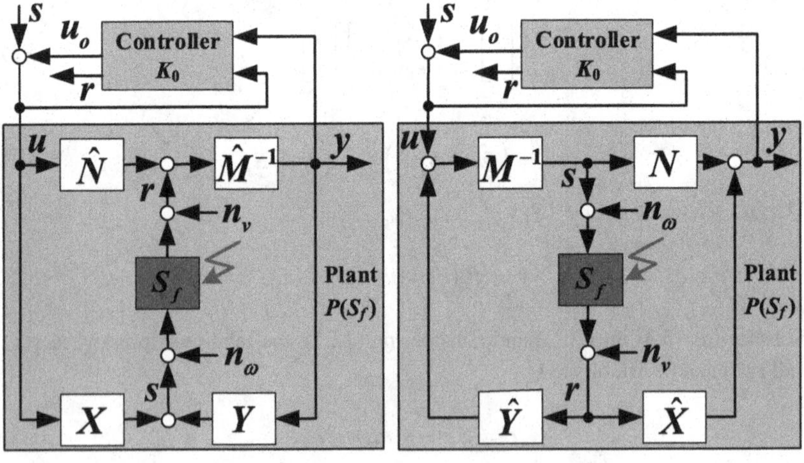

Figure 5.2 LCF and RCF of $\boldsymbol{P}(\boldsymbol{S}_f)$

Lemma 5.1 *Let* $\boldsymbol{L}$ *of the nominal controller* $\boldsymbol{K}_0$ *(2.4) be the Kalman gain. Then* $\boldsymbol{n}_v(k)$ *(5.6), independent of any perturbation* $\boldsymbol{S}_f$*, is Gaussian white noise.*

Proof According to (2.5), (5.6), the dynamics of $\boldsymbol{n}_v(k)$ is

$$\begin{cases} \tilde{\boldsymbol{x}}_n(k+1) = (\boldsymbol{A}-\boldsymbol{L}\boldsymbol{C})\,\tilde{\boldsymbol{x}}_n(k) + (\boldsymbol{B}-\boldsymbol{L}\boldsymbol{D})\,\boldsymbol{\omega}(k) - \boldsymbol{L}\boldsymbol{v}(k), \\ \boldsymbol{n}_v(k) = \boldsymbol{C}\tilde{\boldsymbol{x}}_n(k) + \boldsymbol{D}\boldsymbol{\omega}(k) + \boldsymbol{v}(k), \end{cases} \tag{5.8}$$

where $\tilde{\boldsymbol{x}}_n(k)$ is the state vector. Considering $\boldsymbol{P}$ (3.30) and $\boldsymbol{K}_0$ (2.4), the error dynamics is

$$\tilde{\boldsymbol{x}}(k+1) = (\boldsymbol{A}-\boldsymbol{L}\boldsymbol{C})\,\tilde{\boldsymbol{x}}(k) + (\boldsymbol{B}-\boldsymbol{L}\boldsymbol{D})\,\boldsymbol{\omega}(k) - \boldsymbol{L}\boldsymbol{v}(k), \tag{5.9}$$

where $\tilde{\boldsymbol{x}}(k) = \boldsymbol{x}(k) - \hat{\boldsymbol{x}}(k)$ is the estimation error. Seen from (5.8), (5.9), $\tilde{\boldsymbol{x}}_n(k) = \tilde{\boldsymbol{x}}(k)$ holds. Therefore, if $\boldsymbol{L}$ is designed to be the Kalman gain, $\boldsymbol{n}_v(k)$ is Gaussian white noise. □

It turns out that $\boldsymbol{n}_\omega(k)$ having

$$\boldsymbol{n}_\omega(k) = -\boldsymbol{F}\tilde{\boldsymbol{x}}_n(k) + \boldsymbol{\omega}(k), \tag{5.10}$$

where $\tilde{\boldsymbol{x}}_n$ is governed by (5.8), does not have a Gaussian white property. To find the maximum correlation coefficient between $\boldsymbol{n}_\omega(k+1)$ and $\boldsymbol{n}_\omega(k)$, the following matrix is considered

$$\begin{aligned} \mathcal{K} &= \left(\mathbb{E}(\boldsymbol{n}_\omega(k+1)\boldsymbol{n}_\omega^T(k+1))\right)^{-\frac{1}{2}} \mathbb{E}(\boldsymbol{n}_\omega(k+1)\boldsymbol{n}_\omega^T(k)) \left(\mathbb{E}(\boldsymbol{n}_\omega(k)\boldsymbol{n}_\omega^T(k))\right)^{-\frac{1}{2}} \\ &= \left(\boldsymbol{F}\boldsymbol{\Sigma}_{\tilde{x}_n}\boldsymbol{F}^T + \boldsymbol{\Sigma}_\omega\right)^{-\frac{1}{2}} \left(\boldsymbol{F}\boldsymbol{A}_L\boldsymbol{\Sigma}_{\tilde{x}_n}\boldsymbol{F}^T - \boldsymbol{F}\boldsymbol{B}_L\boldsymbol{\Sigma}_\omega\right) \left(\boldsymbol{F}\boldsymbol{\Sigma}_{\tilde{x}_n}\boldsymbol{F}^T + \boldsymbol{\Sigma}_\omega\right)^{-\frac{1}{2}}. \end{aligned} \tag{5.11}$$

where $\boldsymbol{A}_L = \boldsymbol{A} - \boldsymbol{L}\boldsymbol{C}$, $\boldsymbol{B}_L = \boldsymbol{B} - \boldsymbol{L}\boldsymbol{D}$, and $\boldsymbol{\Sigma}_{\tilde{x}_n}$ is the solution to the following Riccati equation

$$\boldsymbol{\Sigma}_{\tilde{x}_n} = \boldsymbol{A}_L\boldsymbol{\Sigma}_{\tilde{x}_n}\boldsymbol{A}_L^T + \boldsymbol{B}_L\boldsymbol{\Sigma}_\omega\boldsymbol{B}_L^T + \boldsymbol{L}\boldsymbol{\Sigma}_v\boldsymbol{L}^T.$$

By a singular value decomposition of $\mathcal{K}$, we have $\mathcal{K} = \boldsymbol{U}_\mathcal{K}\boldsymbol{S}_\mathcal{K}\boldsymbol{V}_\mathcal{K}^T$, where $\boldsymbol{U}_\mathcal{K}$ and $\boldsymbol{V}_\mathcal{K}$ are orthogonal matrices, and $\boldsymbol{S}_\mathcal{K}$ is a diagonal matrix. Then all diagonal entries of $\boldsymbol{S}_{\kappa}$ are singular values of $\mathcal{K}$, and the maximum singular value of $\mathcal{K}$, denoted as $\sigma_{max}(\mathcal{K})$, is the maximum correlation coefficient.

If the following conditions that

- $\boldsymbol{L}$ is designed to be the Kalman gain to ensure $\boldsymbol{n}_v$ to be Gaussian white according to Theorem 5.1;
- $\sigma_{max}(\mathcal{K}) \ll 1$ holds under the above condition

are valid, then the NAC method developed in Subsection 3.3.2 can be used to tun $\boldsymbol{Q}_r$ for robustness optimization. However, if the above conditions are violated, we will also give some tentative solutions in Subsection 5.2.3 to mitigate the color noise problem.

5.2.2 Conditions of Closed-loop Internal Stability

As an on-policy NAC learning method will be used for the tuning of the auxiliary controller $\boldsymbol{Q}_r$, we study conditions of the closed-loop stability of the pair $\left(\boldsymbol{P}(\boldsymbol{S}_f), \boldsymbol{K}(\boldsymbol{Q}_r)\right)$.

Theorem 5.1 *Consider a feedback control loop consisting of* $\boldsymbol{P}(\boldsymbol{S}_f)$ *(2.21) and* $\boldsymbol{K}(\boldsymbol{Q}_r)$ *(2.15), and a positive constant* ε_s*. For all perturbations* $\boldsymbol{S}_f$ *with* $\|\boldsymbol{S}_f\|_\infty \leq \varepsilon_s$*, the feedback control loop is internally stable if and only if* $\|\boldsymbol{Q}_r\|_\infty < 1/\varepsilon_s$.

Proof According to Theorem 2.6, the stability of the pair $\left(\boldsymbol{P}(\boldsymbol{S}_f), \boldsymbol{K}(\boldsymbol{Q}_r)\right)$ is determined by the pair $\left(\boldsymbol{S}_f, \boldsymbol{Q}_r\right)$. Then Theorem 5.1 can be derived based on Theorem 2.2. □

According to the above theorem, assume that there exists such a constant ε_s for all prescribed perturbations that $\|\boldsymbol{S}_f\|_\infty \leq \varepsilon_s$ holds. Then a conservative stability boundary can be established by

$$\|\boldsymbol{Q}_r\|_\infty < 1/\varepsilon_s. \tag{5.12}$$

Alternatively, according to Theorem 2.3 and 2.4, if the prescribed perturbations are expressed in terms of the left coprime factor uncertainty with $\|[\boldsymbol{\Delta}_{\hat{M}}\ \boldsymbol{\Delta}_{\hat{N}}]\|_\infty \leq \varepsilon_l$

or right coprime factor uncertainty with $\left\|\begin{bmatrix}\Delta_M \\ \Delta_N\end{bmatrix}\right\|_\infty \le \varepsilon_r$ with positive constants ε_l and ε_r, then a conservative stability boundary can be established by having

$$\left\|\begin{bmatrix}-\hat{Y} \\ \hat{X}\end{bmatrix} + \begin{bmatrix}M \\ N\end{bmatrix} Q_r\right\|_\infty < 1/\varepsilon_l, \text{ or } \left\|\begin{bmatrix}Y & X\end{bmatrix} + Q_r \begin{bmatrix}\hat{M} & -\hat{N}\end{bmatrix}\right\|_\infty < 1/\varepsilon_r. \quad (5.13)$$

If ε_l and ε_r are unknown, but the initial nominal controller K_0 satisfies the condition expressed in (5.13) at $Q_r = 0$, then we could check the following condition

$$\begin{aligned} &\left\|\begin{bmatrix}-\hat{Y} \\ \hat{X}\end{bmatrix} + \begin{bmatrix}M \\ N\end{bmatrix} Q_r\right\|_\infty \le \left\|\begin{bmatrix}-\hat{Y} \\ \hat{X}\end{bmatrix}\right\|_\infty, \\ \text{or } &\left\|\begin{bmatrix}Y & X\end{bmatrix} + Q_r \begin{bmatrix}\hat{M} & -\hat{N}\end{bmatrix}\right\|_\infty \le \left\|\begin{bmatrix}Y & X\end{bmatrix}\right\|_\infty \end{aligned} \quad (5.14)$$

to examine the closed-loop stability during online learning of Q_r.

In addition, we propose the following evaluation function to examine the closed-loop stability during online learning of Q_r, that is

$$J_s(j) = \sum_{k=Nj}^{N(j+1)-1} \gamma^{(k-Nj)} \left(s^T(k)\, W_s s(k) + r^T(k)\, W_r r(k)\right), \quad (5.15)$$

where N is a time horizon chosen for one update of Q_r, j represents its jth update. A threshold $J_{s.th}$ can be set as the maximum value of J_s in accord with the intolerable performance of the closed-loop pair (S_f, Q_r). Then, the following decision logic can be used to check the closed-loop stability

$$\begin{cases} J_s(j) \le J_{s.th} \Rightarrow \text{stable} \\ J_s(j) > J_{s.th} \Rightarrow \text{unstable}. \end{cases} \quad (5.16)$$

Note that computing (5.15) online brings some additional benefits. For example, we can decide whether the online learning is correctly performed by observing whether $J_s(j)$ reduces as j increases. In addition, we can decide to stop the learning once $J_s(j)$ stays almost invariant. However, the downside of using (5.15) for stability detection is the large detection delay. In the remainder of our study, we will mainly use (5.14) and (5.15) for stability check.

Based on the above analysis of closed-loop stability of the pair $(\boldsymbol{S}_f, \boldsymbol{Q}_r)$, in the next subsection, we will elaborate the on-policy NAC method for robustness optimization.

5.2.3 Robustness Optimization using NAC

The data-driven robustness optimization using the IOR method is performed in the following three steps.

(A). Parameterization of the Control Policy of $\boldsymbol{Q}_r$
We parameterize $\boldsymbol{Q}_r \in \mathcal{R}_{sp}$ as a general output feedback controller by

$$s(k) = \begin{bmatrix} \boldsymbol{\Theta}_{s.l_s} \\ \boldsymbol{\Theta}_{r.l_s} \end{bmatrix}^T \begin{bmatrix} s_{l_s}(k-1) \\ r_{l_s}(k-1) \end{bmatrix} + \xi(k) = \boldsymbol{\Theta}_s^T z_{sr.l_s}(k-1) + \xi(k), \tag{5.17}$$

where

$$\boldsymbol{\Theta}_s = \begin{bmatrix} \boldsymbol{\Theta}_{s.l_s} \\ \boldsymbol{\Theta}_{r.l_s} \end{bmatrix}, \; z_{sr.l_s}(k-1) = \begin{bmatrix} s_{l_s}(k-1) \\ r_{l_s}(k-1) \end{bmatrix},$$

where $\boldsymbol{\Theta}_{s.l_s}$ and $\boldsymbol{\Theta}_{r.l_s}$ are parameter vectors with proper dimensions. We vectorize $\boldsymbol{\Theta}_s$ with $\boldsymbol{\theta}_s = \text{vec}(\boldsymbol{\Theta}_s)$, such that the optimization of $\boldsymbol{\Theta}_s$ can be performed on $\boldsymbol{\theta}_s$. $s_{l_s}(k-1)$ and $\boldsymbol{r}_{l_s}(k-1)$ are past I/O data of $\boldsymbol{S}_f$ with length l_s, defined as

$$s_{l_s}(k-1) = \begin{bmatrix} s(k-1) \\ s(k-2) \\ \vdots \\ s(k-l_s) \end{bmatrix}, \; \boldsymbol{r}_{l_s}(k-1) = \begin{bmatrix} \boldsymbol{r}(k-1) \\ \boldsymbol{r}(k-2) \\ \vdots \\ \boldsymbol{r}(k-l_s) \end{bmatrix},$$

and $\xi(k) \sim N(\boldsymbol{0}, \boldsymbol{\Sigma}_\xi)$ is Gaussian white noise that is used for persistent excitation. It is clear that a relative large variance $\boldsymbol{\Sigma}_\xi$ can reduce the adverse effect of the color noise $\boldsymbol{n}_w(k)$ on learning, as both of them act on the control input of $\boldsymbol{S}_f$. The control policy of $\boldsymbol{Q}_r$ is parameterized by

$$\pi_{\boldsymbol{\theta}_s}\left(s(k)|z_{sr.l_s}(k-1)\right) = \frac{1}{\sqrt{2\pi|\boldsymbol{\Sigma}_\xi|}} e^{-\frac{1}{2}\left\{\left(s(k)-\boldsymbol{\Theta}_s^T z_{sr.l_s}(k-1)\right)^T \boldsymbol{\Sigma}_\xi^{-1}\left(s(k)-\boldsymbol{\Theta}_s^T z_{sr.l_s}(k-1)\right)\right\}}. \tag{5.18}$$

Here the stochastic policy $\pi_{\theta_s}\left(s(k)|z_{sr.l_s}(k-1)\right)$ denotes the probability of taking the control $s(k)$ conditioned on the current collected data $z_{sr.l_s}(k-1)$.

(B). Parameterization of Value Function, Q-function and Advantage Function
Note that both $\boldsymbol{n}_w(k)$ and $\boldsymbol{n}_v(k)$ can be color noise if $\boldsymbol{L}$ of $\boldsymbol{K}_0$ is not the Kalman gain. But we can modify $\boldsymbol{S}_f$ (5.7) by including dynamics of $\boldsymbol{n}_w(k)$ (5.10) and $\boldsymbol{n}_v(k)$ (5.9), and have an equivalent but higher-order system with white noise. Thus, it is reasonable to assume that a Kalman I/O data model of $\boldsymbol{S}_f$ is always available in spite of color noise. The Kalman state estimation of $\boldsymbol{x}_s$ of $\boldsymbol{S}_f$ (5.7) is denoted as $\boldsymbol{x}_{sK}$. It is then approximated by the I/O data of $\boldsymbol{S}_f$ by

$$\boldsymbol{x}_{sK}(k) \approx \boldsymbol{M}_{sr}^T z_{sr.h_s}(k-1), \tag{5.19}$$

where $\boldsymbol{M}_{sr}$ represents a linear mapping, h_s is the length of the past I/O data. It should be noted that Equation (5.19) holds only if $h_s \gg k_s + n$ holds in the color noise case or if $h_s \gg k_s$ holds in the white noise case. Here we choose a sufficiently large length h_s satisfying $h_s \geq l_s$.

Corollary 5.1 *A value function, defined in (5.15), that evaluates all stabilizing output feedback controllers (5.17) of $\boldsymbol{S}_f$ (5.7) can be approximated by*

$$V\left(z_{sr.h_s}(k-1)\right) = z_{sr.h_s}^T(k-1)\,\boldsymbol{P}_{os} z_{sr.h_s}(k-1) + c_{os}, \tag{5.20}$$

where $\boldsymbol{P}_{os}$ is a constant matrix, and c_{os} is a constant.

Proof The proof is similar to that of Theorem 3.1, and it is thus omitted here. □

According to the above corollary, we parameterize the value function, under the stochastic policy π_{θ_s} (5.18), by

$$V^{\pi_{\theta_s}}\left(z_{sr.h_s}(k-1)\right) = \left[\left(z_{sr.h_s}(k-1) \otimes z_{sr.h_s}(k-1)\right)^T \; 1\right] \boldsymbol{p}_{\theta_s}, \tag{5.21}$$

where $\boldsymbol{p}_{\theta_s}$ is a parameterized column vector. Then the Q-function at k, when a control $s(k)$ is taken, can be described by

$$Q^{\pi_{\theta_s}}\left(z_{sr.h_s}(k-1), s(k)\right) = \boldsymbol{s}^T(k)\,\boldsymbol{W}_s \boldsymbol{s}(k) + \boldsymbol{r}^T(k)\,\boldsymbol{W}_r \boldsymbol{r}(k) + \gamma V^{\pi_{\theta_s}}\left(z_{sr.h_s}(k)\right). \tag{5.22}$$

It is clear that when NAC is used, the advantage function can be parameterized based on (3.58) by

$$A^{\pi_{\theta_s}}\left(z_{sr.h_s}(k-1), s(k)\right) = \left(\nabla_{\theta_s} \ln \pi_{\theta_s}\left(s(k)|z_{sr.l_s}(k-1)\right)\right)^T w_{\theta_s}, \tag{5.23}$$

where w_{θ_s} is a parameterized column vector.

(C). Data-driven Implementation of Q_r

Considering the relationship of value function, Q-function and advantage function described in (3.57), one has

$$A^{\pi_{\theta_s}}\left(z_{sr.h_s}(k-1), s(k)\right) = Q^{\pi_{\theta_s}}\left(z_{sr.h_s}(k-1), s(k)\right) - V^{\pi_{\theta_s}}\left(z_{sr.h_s}(k-1)\right). \tag{5.24}$$

The logarithmic derivative of the stochastic policy (5.18) evaluated at $\theta_{s.j}$ is

$$\begin{aligned} &\nabla_{\theta_s} \ln \pi_{\theta_s}\left(s(k)|z_{sr.l_s}(k-1)\right)\Big|_{\theta_s=\theta_{s.j}} \\ &= \left(I_{k_u} \otimes z_{sr.l_s}(k-1)\right) \Sigma_\xi^{-1} \left(s(k) - \Theta_{s.j}^T z_{sr.l_s}(k-1)\right). \end{aligned} \tag{5.25}$$

Substituting (5.21), (5.22) and (5.23) into the corresponding items in (5.24), and considering (5.25), Equation (5.24) can be rewritten in the form of TD error that is

$$\begin{aligned} \delta(k) =& \left[\left(z_{sr.h_s}(k-1) \otimes z_{sr.h_s}(k-1)\right)^T \ 1\right] p_{\theta_{s.j}} - s^T(k) W_s s(k) - r^T(k) W_r r(k) \\ &- \gamma \left[\left(z_{sr.h_s}(k) \otimes z_{sr.h_s}(k)\right)^T \ 1\right] p_{\theta_{s.j}} \\ &+ \left(\left(I_{k_u} \otimes z_{sr.l_s}(k-1)\right) \Sigma_\xi^{-1} \left(s(k) - \Theta_{s.j}^T z_{sr.l_s}(k-1)\right)\right)^T w_{\theta_{s.j}}. \end{aligned} \tag{5.26}$$

Minimizing this error using LS or SGD, parameters $p_{\theta_{s.j}}$ and $w_{\theta_{s.j}}$ can be online identified with a sequence of s and r data collected under the stochastic policy $\pi_{\theta_{s.j}}$. Subsequently, the policy parameter vector can be optimized, according to (3.64), by

$$\theta_{s.j+1} = \theta_{s.j} - \alpha_n w_{\theta_{s.j}}, \tag{5.27}$$

where α_n is a positive constant step-size.

To sum up, the data-driven algorithm for optimizing the parameter vector θ_s of the auxiliary controller Q_r is given in Algorithm 5.1.

Algorithm 5.1 On-policy NAC: Data-driven Optimization of the Controller $\boldsymbol{Q}_r$

1: **Initialization**:

- Choose a sufficiently large length h_s of the past $\boldsymbol{s}$ and $\boldsymbol{r}$ data for the parameterization of the value function (5.21);
- Choose a length l_s ($l_s \leq h_s$) of past $\boldsymbol{s}$ and $\boldsymbol{r}$ data for the parameterization of the controller $\boldsymbol{Q}_r$ (5.17);
- Select an initial control parameter vector $\boldsymbol{\theta}_{s.0}$ and a positive constant step-size α_n.

for $j = 0, 1, \ldots$

2: **Repeat**

3: **Policy evaluation**: Identify $\boldsymbol{p}_{\boldsymbol{\theta}_{s.j}}$ and $\boldsymbol{w}_{\boldsymbol{\theta}_{s.j}}$ using LS or SGD by minimizing the TD error (5.26) with a sequence of $\boldsymbol{s}$ and $\boldsymbol{r}$, and control parameter $\boldsymbol{\theta}_{s.j}$. Calculate $J_s(j)$ using (5.15) and check closed-loop stability using (5.16).

4: **Policy improvement**: Compute $\boldsymbol{\theta}_{s.j+1}$ using (5.27), and check closed-loop stability by (5.14). Then update $\boldsymbol{\theta}_{s.j}$ to $\boldsymbol{\theta}_{s.j+1}$.

5: **Until** $\boldsymbol{\theta}_s$ converges or some end conditions are satisfied.

5.2.4 Simulation Results

We consider a stochastic plant P that is extended from the SISO plant P_0 in Subsection 3.2.5, described in the following continuous-time state-space form

$$\dot{\boldsymbol{x}} = \boldsymbol{A}\boldsymbol{x} + \boldsymbol{b}(u + \omega), \quad y(k) = \boldsymbol{c}\boldsymbol{x} + v,$$

where $\omega \sim N(0, 10^{-2})$, $v \sim N(0, 10^{-3})$ are actuator and sensor noise, respectively. A nominal controller K_0 (2.4) is designed with a state feedback gain $\boldsymbol{F}^*$ as given in Subsection 3.25 and a Kalman gain $\boldsymbol{L}^*$ calculated based on Equation (3.40). They are

$$\boldsymbol{F}^* = \left[-1.9932\ -2.3266\right], \boldsymbol{L}^* = \left[0.0019\ 0.0171\right]^T.$$

Suppose the perturbed stochastic plant $P(S_f)$ has the following continuous-time state-space representation

$$\dot{\boldsymbol{x}}_{\boldsymbol{P}} = \boldsymbol{A}_{\boldsymbol{P}}\boldsymbol{x}_{\boldsymbol{P}} + \boldsymbol{b}_{\boldsymbol{P}}(u + \omega), \quad y = \boldsymbol{c}_{\boldsymbol{P}}\boldsymbol{x}_{\boldsymbol{P}} + v,$$

with plant matrices given in Subsection 4.2.4. The zero input responses of systems before and after perturbations at the same initial condition are compared and shown in Fig. 5.3. It is clear that the performance of the perturbed system degrades. In order

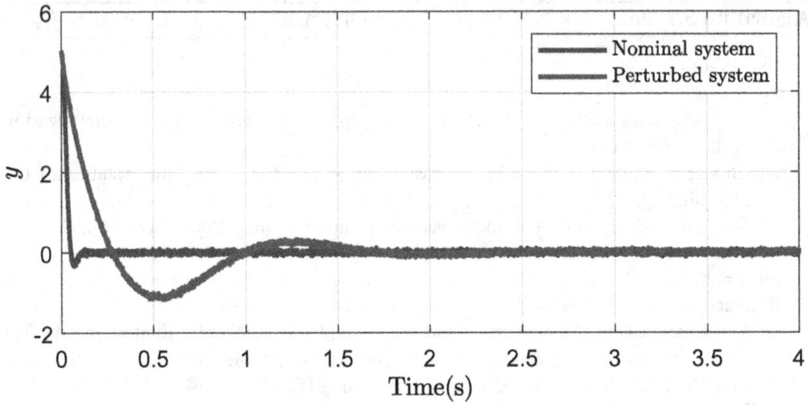

Figure 5.3 Comparison of zero input response of stochastic systems before and after perturbations

to improve the system performance, the proposed IOR strategy with the on-policy NAC learning method is applied.

Before we apply the learning method, the conditions given the Subsection 5.2.1 should be first checked. It can be computed that $\sigma_{max}(\mathcal{K}) = 0.1099$ is relatively smaller than 1, which suggests that on-policy NAC learning can be applied. Then, We parameterize the controller Q_r using the past s and r data with a length $l_s = 2$, and the value function (5.21) with a length $h_s = 8$. The weighting matrices of the value function (5.1) is chosen the same as given in Subsection 4.2.4.

Fig. 5.4 shows the change of control parameters of Q_r during online learning. The cost J_s (5.16) is computed online to ensure the stability of the closed-loop, whose change over the number of the policy updates is shown in Fig. 5.5. After 307 policy updates, the learning procedure is stopped. The learned Q_r (4.19) with system matrices

$$A_q = \begin{bmatrix} -0.1762 & 1 \\ -0.1677 & 0 \end{bmatrix}, b_q = \begin{bmatrix} 16.6938 \\ 6.5160 \end{bmatrix}, c_q = \begin{bmatrix} 1 \\ 0 \end{bmatrix}^T$$

is placed back into the original closed-loop system. Run the closed-loop system again. Its zero input response and the one of the original closed-loop system without Q_r are compared and shown in Fig. 5.6. From the Nyquist diagrams in Figure 5.6, it can be seen that with Q_r the perturbed stochastic system has significantly less undershoot. In addition, from Fig. 5.7, it can be observed that both the gain margin

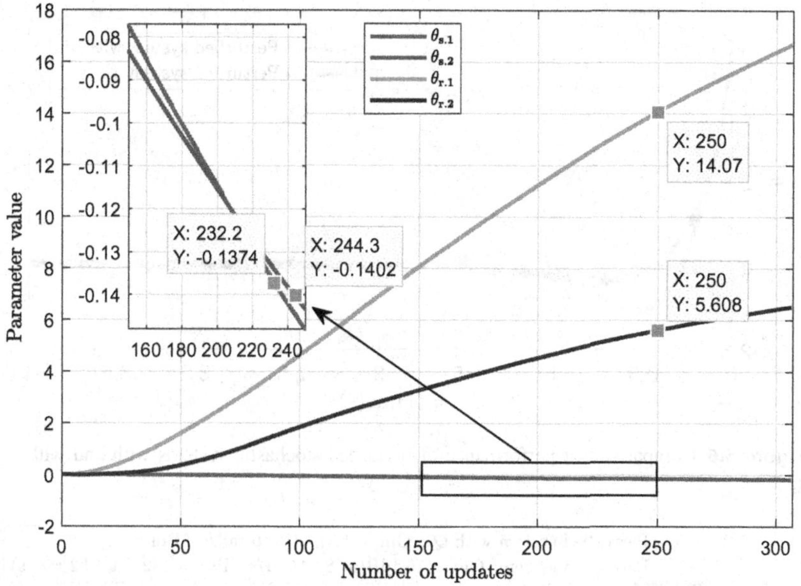

Figure 5.4 Parameters of Q_r during online learning

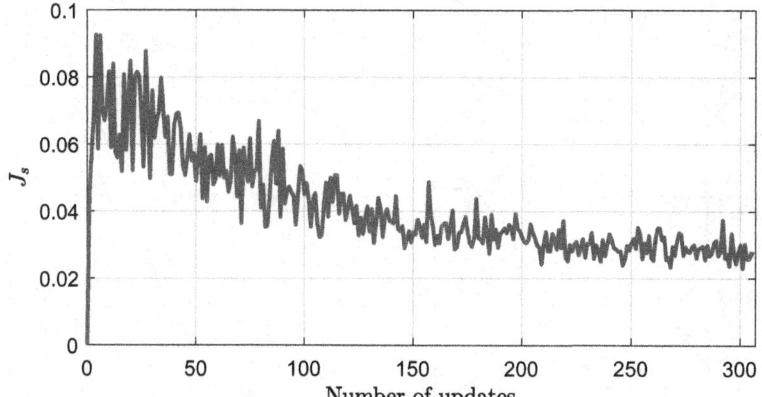

Figure 5.5 Cost J_s during online learning

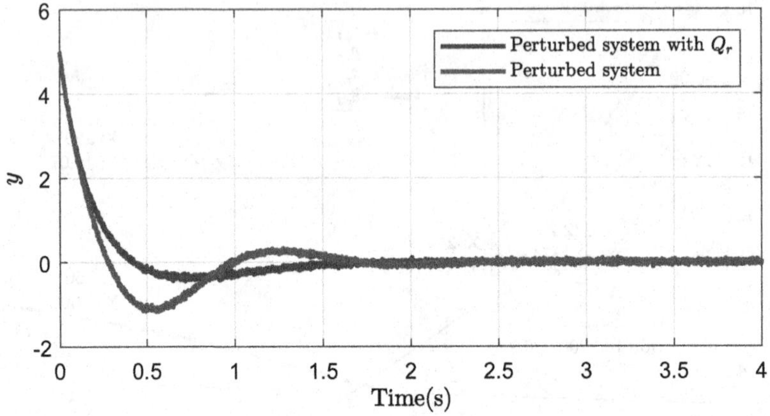

Figure 5.6 Comparison of performance of perturbed stochastic systems with and without Q_r

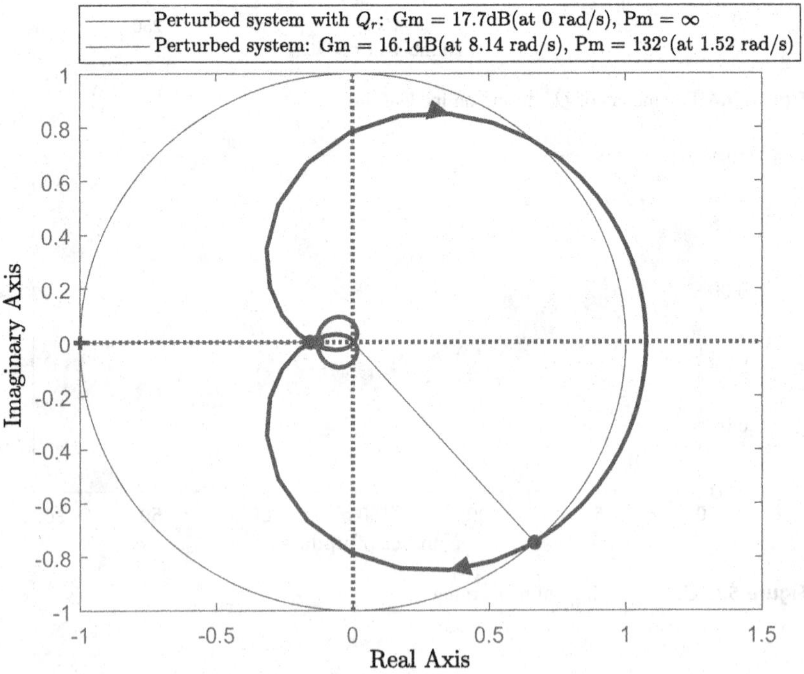

Figure 5.7 Comparison of robustness of perturbed stochastic systems with and without Q_r

and the phase margin of the perturbed system with Q_r are larger than that of the one without Q_r. All of them suggest that the robustness and performance of the perturbed system are improved with the learned Q_r. This demonstrates the effectiveness of the proposed IOR strategy.

5.3 Performance Optimization using a Prescribed Performance Index

In this section, we develop an online controller reconfiguration mechanism to optimize system performance according to the performance index (3.33), based on which $\boldsymbol{K}_0$ is designed. To solve the performance optimization problem, we adopt an auxiliary controller $\boldsymbol{Q}_r$ that feeds back the residual signal back to the control input as the solution as in the previous section.

5.3.1 Efficient NAC Learning using Both Nominal Plant Model and Data

We adopt $\boldsymbol{Q}_r \in \mathcal{R}_{sp}$ as a general dynamic controller in the form described by (5.17). The control policy $\pi_{\theta_s}\left(s(k)|z_{sr.l_s}(k-1)\right)$ of $\boldsymbol{Q}_r$ is expressed in (5.18). We will then apply NAC learning to optimize the parameters of $\boldsymbol{Q}_r$ by minimizing the prescribed value function (3.33). To efficiently perform NAC learning, a parameterized value function that enables the evaluation of all $\boldsymbol{Q}_r$ in the form described by (5.17) is the most critical.

It should be noted that the new controller $\boldsymbol{K}(\boldsymbol{Q}_r) \in \mathcal{R}_{sp}$ is, essentially, an output feedback controller of the plant $\boldsymbol{P}(\boldsymbol{S}_f)$. Therefore, the value function provided by Theorem 3.1 can be directly applied for the evaluation of $\boldsymbol{K}(\boldsymbol{Q}_r)$ using the past I/O data of the plant $\boldsymbol{P}(\boldsymbol{S}_f)$. However, the downside of using the value function is that its construction requires a length of the plant I/O data that is several times as large as the order of $\boldsymbol{P}(\boldsymbol{S}_f)$. This can render online performance optimization less data-efficient, particularly, for MIMO plants.

In order to deal with this, we give an alternative value function below in Corollary 5.2, which is constructed more efficiently using the information of both model of the nominal plant $\boldsymbol{P}_0$ and data. The key idea is to describe the dynamics of the plant $\boldsymbol{P}(\boldsymbol{S}_f)$ using its observer-based I/O data model that is derived based on Equation (2.23) and Equation (5.5) and shown in Fig. 5.8. As in the previous section, we represent the state $\boldsymbol{x}_s$ of $\boldsymbol{S}_f$ by its Kalman state estimation $\boldsymbol{x}_{sK}$ and approximate it by the I/O data of $\boldsymbol{S}_f$ by Equation (5.19).

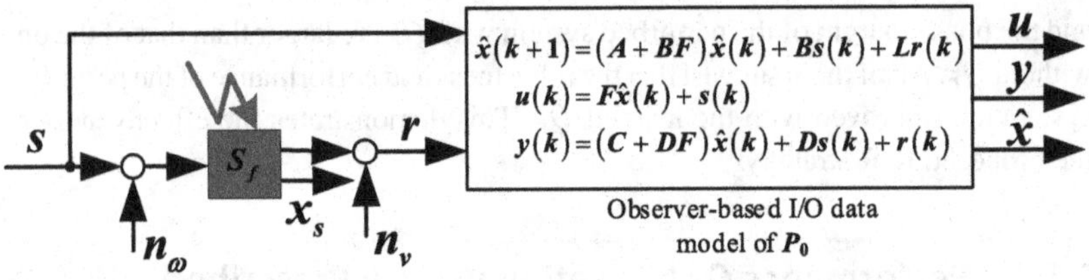

Figure 5.8 Observer-based I/O data model of the stochastic plant $P(S_f)$

Corollary 5.2 *A value function, defined in (3.33), that evaluates all stabilizing output feedback controllers $K(Q_r)$ (2.15) of $P(S_f)$ (2.21) can be approximated by*

$$V\left(\hat{x}(k), z_{sr.h_s}(k-1)\right) = \begin{bmatrix} \hat{x}(k) \\ z_{sr.h_s}(k-1) \end{bmatrix}^T P_s \begin{bmatrix} \hat{x}(k) \\ z_{sr.h_s}(k-1) \end{bmatrix} + c_s, \quad (5.28)$$

where P_s is a constant matrix, and c_s is a constant.

Proof The dynamics of $P(S_f)$ is determined by both $\hat{x}(k)$ and the states of S_f. And the states of S_f are represented by its I/O data in (5.19). Based on these conditions, Corollary 5.2 can be derived in the same manner as Theorem 3.1. □

It is clear that the value function given in Corollary 5.2 makes use of the existing plant model, which is shown by the use of $\hat{x}$. Its construction requires a length of I/O data of S_f that is several times as large as the order of S_f. It should be noted that the order of S_f is often much smaller than that of $P(S_f)$ [68]. Therefore the value function described in (5.28), compared with the one given in Theorem 3.1, is more compact and data-efficient. Due to its data-efficiency, in the next subsection, the value function (5.28) is used for the data-driven performance optimization using NAC.

5.3.2 Data-driven Performance Optimization

To guarantee the closed-loop stability while Q_r is optimized online, all conditions (5.12), (5.13), (5.14) can be adopted. Alternatively, an evaluation function for examining closed-loop stability can be expressed by

$$J_u(j) = \sum_{k=Nj}^{N(j+1)-1} \gamma^{(k-Nj)} \left(\boldsymbol{u}^T(k) \boldsymbol{R} \boldsymbol{u}(k) + \boldsymbol{y}^T(k) \boldsymbol{Q} \boldsymbol{y}(k) \right), \tag{5.29}$$

where N is a time horizon chosen for one update of $\boldsymbol{Q}_r$, j represents its jth update. A threshold $J_{u.th}$ can be set as the maximum value of J_u in accord with the intolerable performance of the system consisting of $\boldsymbol{P}(\boldsymbol{S}_f)$ and $\boldsymbol{K}(\boldsymbol{Q}_r)$. Then, the following decision logic can be used to check the closed-loop stability

$$\begin{cases} J_u(j) \leq J_{u.th} \Rightarrow \text{stable} \\ J_u(j) > J_{u.th} \Rightarrow \text{unstable}. \end{cases} \tag{5.30}$$

In what follows, we elaborate the data-driven implementation of the performance index based optimization method in the following two steps.

(A). Parameterization of Value Function, Q-function and Advantage Function
We parameterize the value function (5.28), under the stochastic policy π_{θ_s} (5.18), by

$$V^{\pi_{\theta_s}}\left(\hat{\boldsymbol{x}}(k), z_{sr.h_s}(k-1)\right) = \left[\left(\begin{bmatrix} \hat{\boldsymbol{x}}(k) \\ z_{sr.h_s}(k-1) \end{bmatrix} \otimes \begin{bmatrix} \hat{\boldsymbol{x}}(k) \\ z_{sr.h_s}(k-1) \end{bmatrix} \right)^T \; 1 \right] \boldsymbol{p}_{\theta_s}, \tag{5.31}$$

where $\boldsymbol{p}_{\theta_s}$ is a parameterized column vector. Then the Q-function at k, when a control $\boldsymbol{s}(k)$ is taken, can be described by

$$\begin{aligned} Q^{\pi_{\theta_s}}\left(\hat{\boldsymbol{x}}(k), z_{sr.h_s}(k-1), \boldsymbol{s}(k)\right) =& \boldsymbol{u}^T(k) \boldsymbol{R} \boldsymbol{u}(k) + \boldsymbol{y}^T(k) \boldsymbol{Q} \boldsymbol{y}(k) \\ &+ \gamma V^{\pi_{\theta_s}}\left(\hat{\boldsymbol{x}}(k), z_{sr.h_s}(k)\right), \end{aligned} \tag{5.32}$$

and the advantage function can be parameterized by

$$A^{\pi_{\theta_s}}\left(\hat{\boldsymbol{x}}(k), z_{sr.h_s}(k-1), \boldsymbol{s}(k)\right) = \left(\nabla_{\theta_s} \ln \pi_{\theta_s}\left(\boldsymbol{s}(k) | z_{sr.l_s}(k-1)\right)\right)^T \boldsymbol{w}_{\theta_s}, \tag{5.33}$$

where $\boldsymbol{w}_{\theta_s}$ is a parameterized column vector.

(B). Data-driven Implementation of $\boldsymbol{Q}_r$
Considering the relationship of value function, Q-function and advantage function described in (3.57), one has

$$\begin{aligned} A^{\pi_{\theta_s}}\left(\hat{\boldsymbol{x}}(k), z_{sr.h_s}(k-1), \boldsymbol{s}(k)\right) =& Q^{\pi_{\theta_s}}\left(\hat{\boldsymbol{x}}(k), z_{sr.h_s}(k-1), \boldsymbol{s}(k)\right) \\ &- V^{\pi_{\theta_s}}\left(\hat{\boldsymbol{x}}(k), z_{sr.h_s}(k-1)\right). \end{aligned} \tag{5.34}$$

The logarithmic derivative of $\pi_{\theta_s}\left(s(k)|z_{sr.l_s}(k-1)\right)$ evaluated at $\boldsymbol{\theta}_{s.j}$ is expressed by (5.25). Substituting (5.31), (5.32) and (5.33) into the corresponding items in (5.34), and considering (5.25), Equation (5.34) can be rewritten in the form of TD error that is

$$\begin{aligned}\delta(k)=&\left[\left(\begin{bmatrix}\hat{\boldsymbol{x}}(k)\\ z_{sr.h_s}(k-1)\end{bmatrix}\otimes\begin{bmatrix}\hat{\boldsymbol{x}}(k)\\ z_{sr.h_s}(k-1)\end{bmatrix}\right)^T 1\right]\boldsymbol{p}_{\boldsymbol{\theta}_{s.j}}-\boldsymbol{u}^T(k)\,\boldsymbol{R}\boldsymbol{u}(k)\\&-\boldsymbol{y}^T(k)\,\boldsymbol{Q}\boldsymbol{y}(k)-\gamma\left[\left(\begin{bmatrix}\hat{\boldsymbol{x}}(k+1)\\ z_{sr.h_s}(k)\end{bmatrix}\otimes\begin{bmatrix}\hat{\boldsymbol{x}}(k+1)\\ z_{sr.h_s}(k)\end{bmatrix}\right)^T 1\right]\boldsymbol{p}_{\boldsymbol{\theta}_{s.j}}\\&+\left(\boldsymbol{I}_{k_u}\otimes z_{sr.l_s}(k-1)\right)\boldsymbol{\Sigma}_{\xi}^{-1}\left(s(k)-\boldsymbol{\Theta}_{s.j}^T z_{sr.l_s}(k-1)\right)^T\boldsymbol{w}_{\boldsymbol{\theta}_{s.j}}.\end{aligned} \tag{5.35}$$

Minimizing this error using LS or SGD, parameters $\boldsymbol{p}_{\boldsymbol{\theta}_{s.j}}$ and $\boldsymbol{w}_{\boldsymbol{\theta}_{s.j}}$ can be online identified with a sequence of $\hat{\boldsymbol{x}}$, $\boldsymbol{s}$, $\boldsymbol{r}$, and $\boldsymbol{u}$ and $\boldsymbol{y}$ data collected under the stochastic policy $\pi_{\boldsymbol{\theta}_{s.j}}$. Subsequently, the policy parameter vector can be optimized, according to (3.64), by

$$\boldsymbol{\theta}_{s.j+1}=\boldsymbol{\theta}_{s.j}-\alpha_n\boldsymbol{w}_{\boldsymbol{\theta}_{s.j}}, \tag{5.36}$$

where α_n is a positive constant step-size.

To sum up, the data-driven algorithm for optimizing $\boldsymbol{\theta}_s$ of the auxiliary controller $\boldsymbol{Q}_r$ is given in Algorithm 5.2. The corresponding diagram is shown in Fig. 5.9.

Algorithm 5.2 On-policy NAC: Data-driven Optimization of the Controller $\boldsymbol{Q}_r$

1: **Initialization**:

- Choose a sufficiently large length h_s of the past $\boldsymbol{s}$ and $\boldsymbol{r}$ data for the parameterization of the value function (5.21);
- Choose a length l_s ($l_s \leq h_s$) of past $\boldsymbol{s}$ and $\boldsymbol{r}$ data for the parameterization of the controller $\boldsymbol{Q}_r$ (5.17);
- Select an initial control parameter vector $\boldsymbol{\theta}_{s.0}$ and a positive constant step-size α_n.

for $j=0,1,\ldots$

2: **Repeat**

3: **Policy evaluation**: Identify $\boldsymbol{p}_{\boldsymbol{\theta}_{s.j}}$ and $\boldsymbol{w}_{\boldsymbol{\theta}_{s.j}}$ using LS or SGD by minimizing the TD error (5.35) with a sequence of $\hat{\boldsymbol{x}}$, $\boldsymbol{s}$, $\boldsymbol{r}$, $\boldsymbol{u}$, $\boldsymbol{y}$, and the control parameter $\boldsymbol{\theta}_{s.j}$. Calculate $J_u(j)$ using (5.29) and check closed-loop stability using (5.30).

4: **Policy improvement**: Compute $\boldsymbol{\theta}_{s.j+1}$ using (5.36), and check closed-loop stability by (5.14). Then update $\boldsymbol{\theta}_{s.j}$ to $\boldsymbol{\theta}_{s.j+1}$.

5: **Until** $\boldsymbol{\theta}_s$ converges or some end conditions are satisfied.

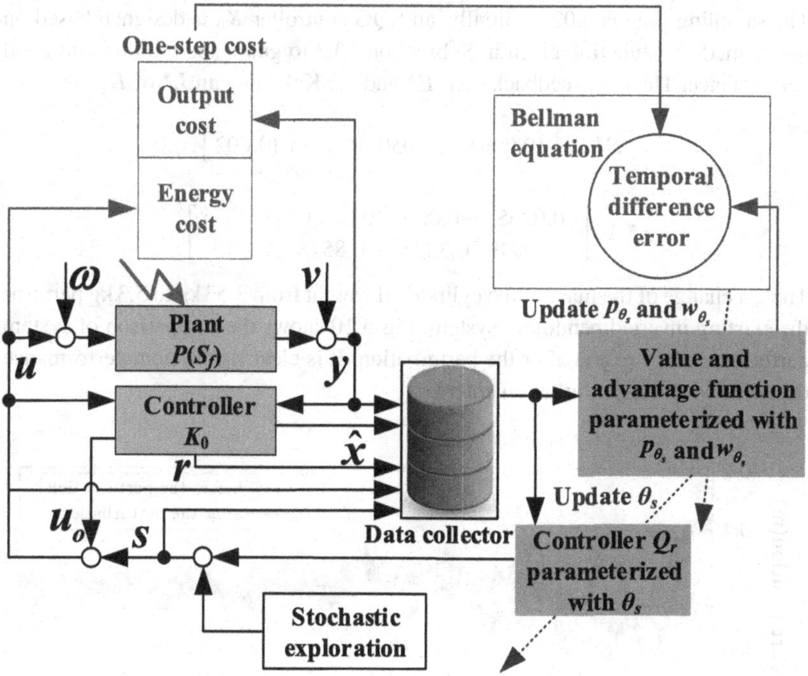

Figure 5.9 Data-driven performance optimization of Q_r using NAC

5.3.3 Simulation Results

In this subsection, a simulation study on an inverted pendulum system is used to show the feasibility of the proposed performance index based optimization approach. The system has the following continuous-time state-space representation form

$$\dot{x} = Ax + b(u + \omega), \quad y = Cx + v,$$

where system matrices are the same as those of the inverted pendulum given in Subsection 4.3.3, and ω and v are actuator and sensor noises having

$$\omega \sim N(0, 0.01),\ v \sim N\left(\begin{bmatrix} 0 \\ 0 \end{bmatrix}, \begin{bmatrix} 10^{-4} & 0 \\ 0 & 10^{-6} \end{bmatrix}\right).$$

The sampling time is 0.02 s. Initially, an LQG controller $\boldsymbol{K}_0$ is designed based on the defined cost function given in Subsection 4.3.3 to guarantee stability and good performance. The state feedback gain $\boldsymbol{F}^*$ and the Kalman gain $\boldsymbol{L}^*$ of $\boldsymbol{K}_0$ are

$$\boldsymbol{F}^* = \begin{bmatrix} 19.9991 & 92.2050 & 20.6656 & 19.692 \end{bmatrix},$$

$$\boldsymbol{L}^* = \begin{bmatrix} 0.0265 & -0.0009 & 0.0183 & -0.0036 \\ -0.0949 & 0.3225 & -0.8518 & 2.3017 \end{bmatrix}^T.$$

Then, a change of the mass of the cylindrical weight from 3.53 kg to 5.3 kg perturbs the existing inverted pendulum system. Fig. 5.10 shows the comparison of system performance before and after the perturbation. It is clear that system performance degrades when the perturbation enters.

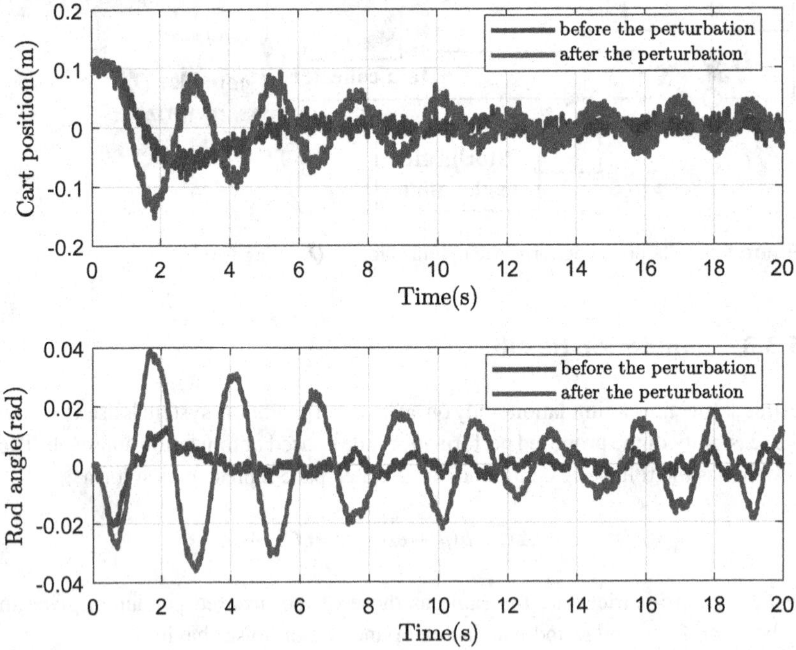

Figure 5.10 Comparison of performance of systems before and after the perturbation

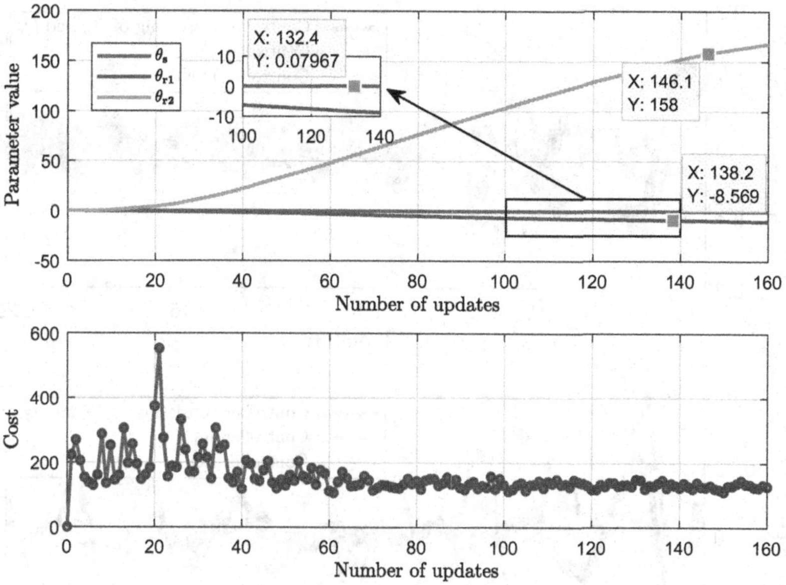

Figure 5.11 Parameters of Q_r and cost J_u during online learning

In order to enhance the performance, a first-order $\boldsymbol{Q_r}$ (5.17) is introduced into the system, whose parameter $\boldsymbol{\Theta}$ is online learned using the above NAC policy gradient method. The length of the past I/O data, s and r, used for the parameterization of the value function (5.21) is set to 4. Fig. 5.11 shows the change of the parameters of $\boldsymbol{Q_r}$ and cost during online learning. The 'Cost' marking the y-axis of the lower plot of Fig. 5.11 is computed by Equation (5.29), where N is chosen to be 460 that is equal to 9.2 s. Inferred from the gradual decreasing cost in Fig. 5.11, the updating directions of the parameters of $\boldsymbol{Q_r}$ are correct.

Finally, after 160 policy updates, the learning procedure is stopped. The learned $\boldsymbol{Q_r}$ with parameter $\boldsymbol{\Theta} = \left[0.1022\ -9.992\ 168.8\right]^T$ is placed back into the system. Fig. 5.12 shows the comparison of performance of systems consisting of the perturbed plant $\boldsymbol{P}(\boldsymbol{S_f})$ and different controllers. It can be seen from Fig. 5.12 that, with the learned $\boldsymbol{Q_r}$ the system shows considerably less overshoot and settling time for both the cart position and the rod angle. This demonstrates the effectiveness of the proposed performance index based optimization method. However, it can also be observed that the system consisting of $\boldsymbol{P}(\boldsymbol{S_f})$ and the LQG controller has a better

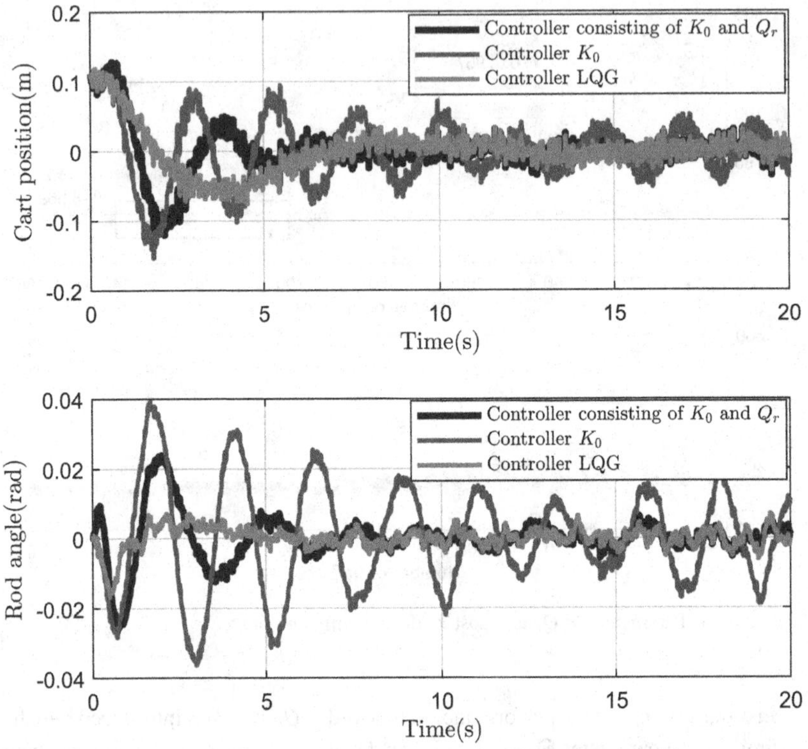

Figure 5.12 Comparison of performance of systems consisting of $P(S_f)$ and different controllers

performance. This shows a drawback of the proposed method that it can only learn a suboptimal controller.

5.4 Performance Optimization of Plants with General Output Feedback Controllers

In the previous studies, we have investigated methods for robustness and performance optimization of feedback control systems. They can be applied to plants with feedback controllers, where a residual vector is available. However, there are some cases where a residual vector of a feedback controller is inaccessible, e.g. a

PID controller. Motivated by this, in this section, we are dedicated to developing approaches for performance optimization of plants with general output feedback controllers.

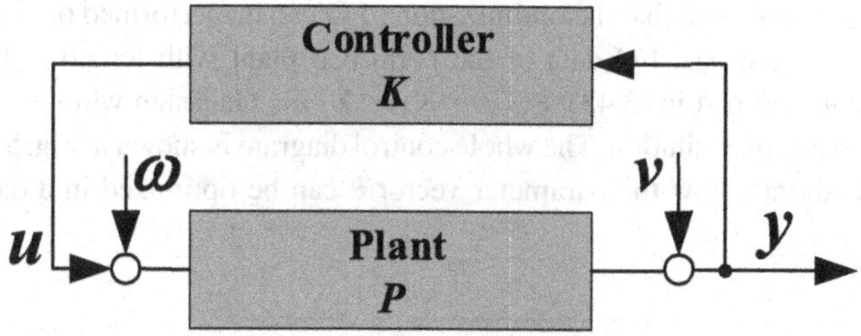

Figure 5.13 Configuration of the system with a general feedback controller

5.4.1 Problem Formulation

We deal with the same stochastic plant $\boldsymbol{P}$ (3.30) as in the previous section, and assume that an initial output feedback controller $\boldsymbol{K} \in \mathcal{R}_p$ is designed to guarantee stability and good performance based on the performance index defined in (3.33). Here the controller $\boldsymbol{K}$ is described by

$$\boldsymbol{K}: \begin{cases} \boldsymbol{x}_c(k+1) = \boldsymbol{A}_c\boldsymbol{x}_c(k) + \boldsymbol{B}_c\boldsymbol{y}(k), \\ \boldsymbol{u}(k) = \boldsymbol{C}_c\boldsymbol{x}_c(k) + \boldsymbol{D}_c\boldsymbol{y}(k), \end{cases} \tag{5.37}$$

where $\boldsymbol{x}_c \in \mathbb{R}^{k_c}$ is the state vector assumed to be measurable. $\boldsymbol{A}_c$, $\boldsymbol{B}_c$, $\boldsymbol{C}_c$, and $\boldsymbol{D}_c$ are constant real matrices with proper dimensions. The configuration is shown in Fig. 5.13.

Assume that perturbations happen in the plant $\boldsymbol{P}$ degrade only system performance and have no influence on stability. The perturbed stochastic plant is denoted as $\boldsymbol{P}_p$. Our goal here is to develop an online controller reconfiguration mechanism to recover system performance according to the prescribed value function (3.33).

To this end, we introduce an auxiliary controller $\boldsymbol{K}_A \in \mathcal{R}_{sp}$ with the following structure to the system

$$\boldsymbol{u}_A(k) = \begin{bmatrix} \boldsymbol{\Theta}_{u.l} \\ \boldsymbol{\Theta}_{y.l} \end{bmatrix}^T \begin{bmatrix} \boldsymbol{u}_l(k-1) \\ \boldsymbol{y}_l(k-1) \end{bmatrix} + \boldsymbol{\xi}(k) = \boldsymbol{\Theta}^T \boldsymbol{z}_{uy.l}(k-1) + \boldsymbol{\xi}(k), \tag{5.38}$$

where

$$\boldsymbol{\Theta} = \begin{bmatrix} \boldsymbol{\Theta}_{u.l} \\ \boldsymbol{\Theta}_{y.l} \end{bmatrix}, \; z_{uy.l}(k-1) = \begin{bmatrix} \boldsymbol{u}_l(k-1) \\ y_l(k-1) \end{bmatrix},$$

where $\boldsymbol{\Theta}_{u.l}$ and $\boldsymbol{\Theta}_{y.l}$ are parameter vectors with proper dimensions. We vectorize $\boldsymbol{\Theta}$ with $\boldsymbol{\theta} = \text{vec}(\boldsymbol{\Theta})$, such that the optimization of $\boldsymbol{\Theta}$ can be performed on $\boldsymbol{\theta}$. $\boldsymbol{u}_l(k-1)$ and $y_l(k-1)$ are past I/O data of the perturbed plant with length l. Their data structures are defined in (3.43). $\boldsymbol{\xi}(k) \sim N(\boldsymbol{0}, \; \boldsymbol{\Sigma}_{\xi})$ is Gaussian white noise that is used for persistent excitation. The whole control diagram is shown in Fig. 5.14. Next, we shall elaborate how the parameter vector $\boldsymbol{\theta}$ can be optimized in a data-driven manner.

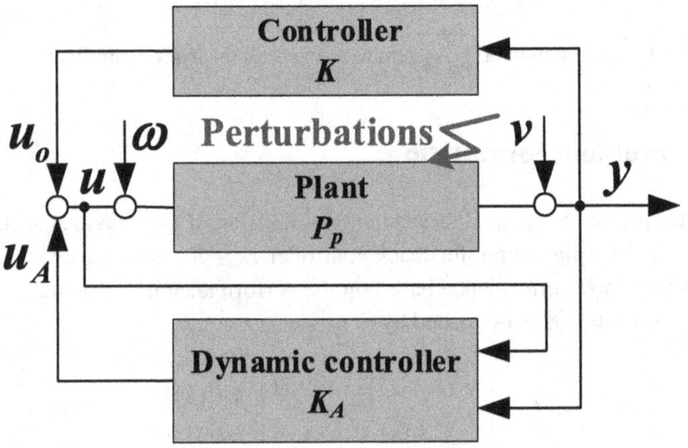

Figure 5.14 Performance optimization with an auxiliary controller $\boldsymbol{K_A}$

5.4.2 Performance Optimization using NAC

We parameterize the control policy of $\boldsymbol{K_A}$ as

$$\pi_{\theta}\left(\boldsymbol{u}_A(k)|z_{uy.l}(k-1)\right) = \frac{1}{\sqrt{2\pi|\boldsymbol{\Sigma}_{\xi}|}} e^{-\frac{1}{2}\left\{\left(\boldsymbol{u}_A(k)-\boldsymbol{\Theta}^T z_{uy.l}(k-1)\right)^T \sigma_{\xi}^{-2}\left(\boldsymbol{u}_A(k)-\boldsymbol{\Theta}^T z_{uy.l}(k-1)\right)\right\}}. \tag{5.39}$$

Here the stochastic policy $\pi_{\theta}\left(\boldsymbol{u}_A(k)|z_{uy.l}(k-1)\right)$ denotes the probability of taking the control $\boldsymbol{u}_A(k)$ conditioned on the current collected I/O data $z_{uy.l}(k-1)$.

(A). Parameterization of the Value Function, Q-function and Advantage Function
As the perturbed plant $\boldsymbol{P_p}$ is unknown, the following Kalman I/O data model is used to equivalently represent its dynamics

$$\begin{cases} \boldsymbol{x_p}(k+1) = \boldsymbol{A_p x_p}(k) + \boldsymbol{B_p u}(k) + \boldsymbol{L_p r_p}(k), \\ \boldsymbol{r_p}(k) = \boldsymbol{y}(k) - \boldsymbol{C_p x_p}(k) - \boldsymbol{D_p u}(k), \end{cases} \tag{5.40}$$

where $\boldsymbol{x_p} \in \mathcal{R}^{k_p}$ is the Kalman state estimate and $\boldsymbol{L_p}$ is the Kalman gain, and $\boldsymbol{r_p}$ is the innovation signal satisfying $\boldsymbol{r_p} \sim N(\boldsymbol{0}, \boldsymbol{\Sigma_r})$, and $\boldsymbol{A_p}$, $\boldsymbol{B_p}$, $\boldsymbol{C_p}$, and $\boldsymbol{D_p}$ are unknown constant real matrices with proper dimensions. It is clear that $\boldsymbol{x_p}(k)$ can also be approximated by I/O data of $\boldsymbol{P_p}$

$$\boldsymbol{x_p}(k) \approx \boldsymbol{M_p^T z_{uy.h}}(k-1), \tag{5.41}$$

where $\boldsymbol{M_p}$ represents a linear mapping, and $h \gg k_p$. Here we choose a large length h satisfying $h \geq l$.

Corollary 5.3 *A value function, defined in (3.33), that evaluates all stabilizing controllers consisting of $\boldsymbol{K}$ (5.37) and $\boldsymbol{K_A}$ (5.38) of the perturbed plant $\boldsymbol{P_p}$ can be approximated by*

$$V\left(\boldsymbol{x_c}(k), \boldsymbol{z_{uy.h}}(k-1)\right) = \begin{bmatrix} \boldsymbol{x_c}(k) \\ \boldsymbol{z_{uy.h}}(k-1) \end{bmatrix}^T \boldsymbol{P_u} \begin{bmatrix} \boldsymbol{x_c}(k) \\ \boldsymbol{z_{uy.h}}(k-1) \end{bmatrix} + c_u, \tag{5.42}$$

where $\boldsymbol{P_u}$ is a constant matrix, and c_u is a constant.

Proof The closed-loop dynamics of $\boldsymbol{P_p}$ and $\boldsymbol{K}$ is determined by both $\boldsymbol{x_c}(k)$ and the states of $\boldsymbol{P_p}$. And the states of $\boldsymbol{P_p}$ are represented by its I/O data in (5.41). Based on these conditions, Corollary 5.3 can be derived in the same manner as Theorem 3.1. □

We parameterize the value function (5.42), under the stochastic policy π_θ (5.39), by

$$V^{\pi_\theta}\left(\boldsymbol{x_c}(k), \boldsymbol{z_{uy.h}}(k-1)\right) = \left[\left(\begin{bmatrix} \boldsymbol{x_c}(k) \\ \boldsymbol{z_{uy.h}}(k-1) \end{bmatrix} \otimes \begin{bmatrix} \boldsymbol{x_c}(k) \\ \boldsymbol{z_{uy.h}}(k-1) \end{bmatrix}\right)^T 1\right] \boldsymbol{p_\theta}, \tag{5.43}$$

where $\boldsymbol{p_\theta}$ is a parameterized column vector. Then the Q-function at k, when a control $\boldsymbol{u_A}(k)$ is taken, can be described by

$$\begin{aligned} Q^{\pi_\theta}\left(\boldsymbol{x}_c(k), \boldsymbol{z}_{uy.h}(k-1), \boldsymbol{u}_A(k)\right) =& \boldsymbol{u}^T(k)\boldsymbol{R}\boldsymbol{u}(k) + \boldsymbol{y}^T(k)\boldsymbol{Q}\boldsymbol{y}(k) \\ &+ \gamma V^{\pi_\theta}\left(\boldsymbol{x}_c(k), \boldsymbol{z}_{uy.h}(k)\right), \end{aligned} \tag{5.44}$$

and the advantage function can be parameterized by

$$A^{\pi_\theta}\left(\boldsymbol{x}_c(k), \boldsymbol{z}_{uy.h}(k-1), \boldsymbol{u}_A(k)\right) = \left(\nabla_\theta \ln \pi_\theta\left(\boldsymbol{u}_A(k)|\boldsymbol{z}_{uy.l}(k-1)\right)\right)^T \boldsymbol{w}_\theta, \tag{5.45}$$

where $\boldsymbol{w}_\theta$ is a parameterized column vector.

(B). Data-driven Implementation of $\boldsymbol{K}_A$
Considering the relationship of value function, Q-function and advantage function described in (3.57), one has

$$\begin{aligned} A^{\pi_\theta}\left(\boldsymbol{x}_c(k), \boldsymbol{z}_{uy.h}(k-1), \boldsymbol{u}_A(k)\right) =& Q^{\pi_\theta}\left(\boldsymbol{x}_c(k), \boldsymbol{z}_{uy.h}(k-1), \boldsymbol{u}_A(k)\right) \\ &- V^{\pi_\theta}\left(\boldsymbol{x}_c(k), \boldsymbol{z}_{uy.h}(k-1)\right). \end{aligned} \tag{5.46}$$

The logarithmic derivative of the stochastic policy (5.39) evaluated at $\boldsymbol{\theta}_j$ is

$$\begin{aligned} &\nabla_\theta \ln \pi_\theta\left(\boldsymbol{u}_A(k)|\boldsymbol{z}_{uy.l}(k-1)\right)\big|_{\theta=\theta_j} \\ &= \left(\boldsymbol{I}_{k_u} \otimes \boldsymbol{z}_{uy.l}(k-1)\right)\boldsymbol{\Sigma}_\xi^{-1}\left(\boldsymbol{u}_A(k) - \boldsymbol{\Theta}_j^T \boldsymbol{z}_{uy.l}(k-1)\right). \end{aligned} \tag{5.47}$$

Substituting (5.43), (5.44) and (5.45) into the corresponding items in (5.46), and considering (5.47), Equation (5.46) can be rewritten in the form of the TD error that is

$$\begin{aligned} \delta(k) =& \left[\left(\begin{bmatrix} \boldsymbol{x}_c(k) \\ \boldsymbol{z}_{uy.h}(k-1) \end{bmatrix} \otimes \begin{bmatrix} \boldsymbol{x}_c(k) \\ \boldsymbol{z}_{uy.h}(k-1) \end{bmatrix}\right)^T \ 1\right] \boldsymbol{p}_{\theta_j} - \boldsymbol{u}^T(k)\boldsymbol{R}\boldsymbol{u}(k) \\ &- \boldsymbol{y}^T(k)\boldsymbol{Q}\boldsymbol{y}(k) - \gamma\left[\left(\begin{bmatrix} \boldsymbol{x}_c(k+1) \\ \boldsymbol{z}_{uy.h}(k) \end{bmatrix} \otimes \begin{bmatrix} \boldsymbol{x}_c(k+1) \\ \boldsymbol{z}_{uy.h}(k) \end{bmatrix}\right)^T \ 1\right] \boldsymbol{p}_{\theta_j} \\ &+ \left(\boldsymbol{I}_{k_u} \otimes \boldsymbol{z}_{uy.l}(k-1)\right)\boldsymbol{\Sigma}_\xi^{-1}\left(\boldsymbol{u}_A(k) - \boldsymbol{\Theta}_j^T \boldsymbol{z}_{uy.l}(k-1)\right)^T \boldsymbol{w}_{\theta_j}. \end{aligned} \tag{5.48}$$

Minimizing this error using LS or SGD, parameters $\boldsymbol{p}_{\theta_j}$ and $\boldsymbol{w}_{\theta_j}$ can be online identified with a sequence of $\boldsymbol{x}_c$, $\boldsymbol{u}$, $\boldsymbol{u}_A$ and $\boldsymbol{y}$ data collected under the stochastic policy π_{θ_j}. Subsequently, the policy parameter vector can be optimized, according to (3.64), by

$$\boldsymbol{\theta}_{j+1} = \boldsymbol{\theta}_j - \alpha_n \boldsymbol{w}_{\theta_j}, \tag{5.49}$$

where α_n is a positive constant step-size.

To sum up, the data-driven algorithm for optimizing the parameter vector $\boldsymbol{\theta}$ of the auxiliary controller $\boldsymbol{K_A}$ is given in Algorithm 5.3. The corresponding diagram is shown in Fig. 5.15.

Algorithm 5.3 On-policy NAC: Data-driven Optimization of the Controller $\boldsymbol{K_A}$

1: **Initialization**:

- Choose a sufficiently large length h of the past $\boldsymbol{u}$ and $\boldsymbol{y}$ data for the parameterization of the value function (5.43);
- Choose a length l ($l \leq h$) of past $\boldsymbol{u}$ and $\boldsymbol{y}$ data for the parameterization of the controller $\boldsymbol{K_A}$ (5.38);
- Select an initial control parameter vector $\boldsymbol{\theta}_0$ and a positive constant step-size α_n.

for $j = 0, 1, \ldots$

2: **Repeat**

3: **Policy evaluation**: Identify $\boldsymbol{p}_{\boldsymbol{\theta}_j}$ and $\boldsymbol{w}_{\boldsymbol{\theta}_j}$ using LS or SGD by minimizing the TD error (5.48) with a sequence of $\boldsymbol{x}_c$, $\boldsymbol{u}$, $\boldsymbol{u}_A$, $\boldsymbol{y}$, and the control parameter $\boldsymbol{\theta}_j$. Calculate $J_u(j)$ using (5.29) and check closed-loop stability using (5.30).

4: **Policy improvement**: Update $\boldsymbol{\theta}_j$ to $\boldsymbol{\theta}_{j+1}$ using (5.49).

5: **Until** $\boldsymbol{\theta}$ converges or some end conditions are satisfied.

Remark 5.1 *The value function given in (5.42) can evaluate not only the auxiliary controller* $\boldsymbol{K_A}$ *with the form (5.38), but also the one with the feedback of states* $\boldsymbol{x_c}$ *of* $\boldsymbol{K}$ *to the control input, e.g.* $\boldsymbol{u_A}(k) = \boldsymbol{\theta_x x_c}(k)$ *where* $\boldsymbol{\theta_x}$ *is a parameter vector to be optimized, or even the one with the combination of the two forms. Thus, we suggest that domain knowledge be used to parameterize* $\boldsymbol{K_A}$ *in order to achieve best possible performance optimization with least parameters.*

5.4.3 Experimental Results

To demonstrate the proposed data-driven method for performance optimization, we apply it to a BLDC motor test rig, seen in Fig. 5.16. It consists of two BLDC motors coupled with a mechanical shaft. The Maxon motor serves as a driving motor and the Nanotec motor acts as a load.

The control objective of a BLDC motor test rig is to achieve good speed tracking performance. To this end, the BLDC motor test rig is controlled in the dq coordinate system using a vector control strategy. Initially, two proportional-integral (PI)

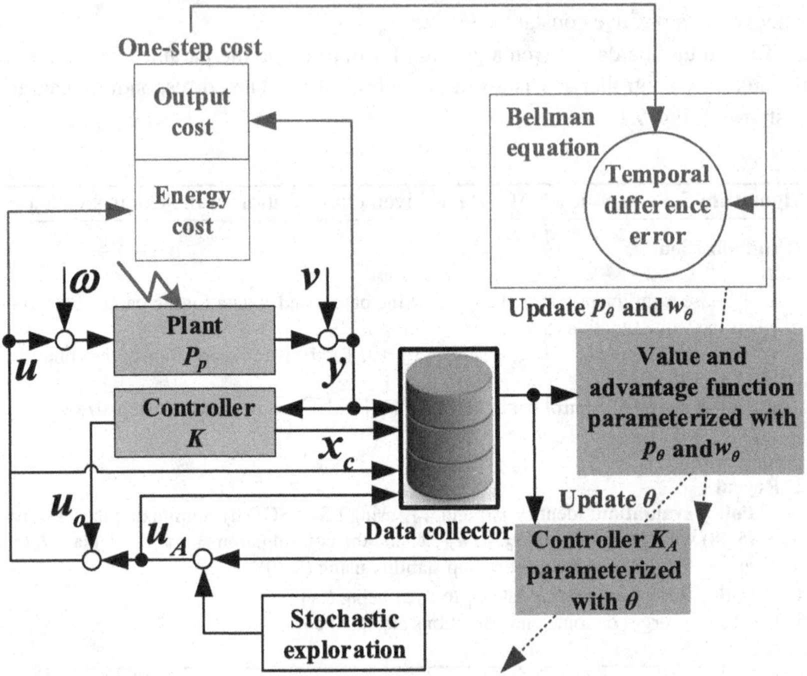

Figure 5.15 Data-driven performance optimization of K_A using NAC

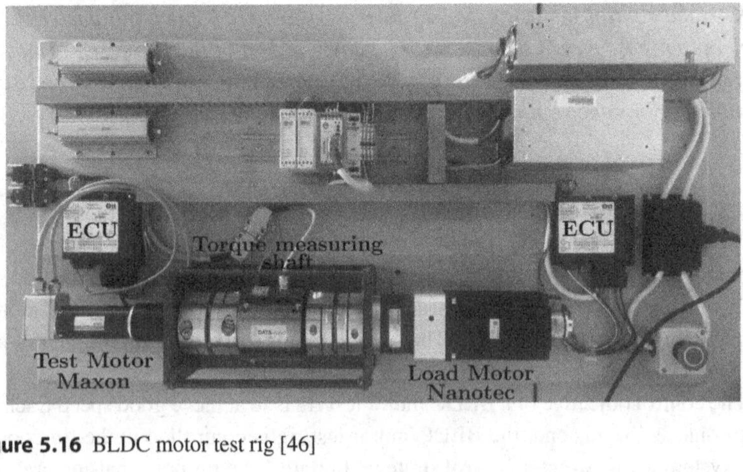

Figure 5.16 BLDC motor test rig [46]

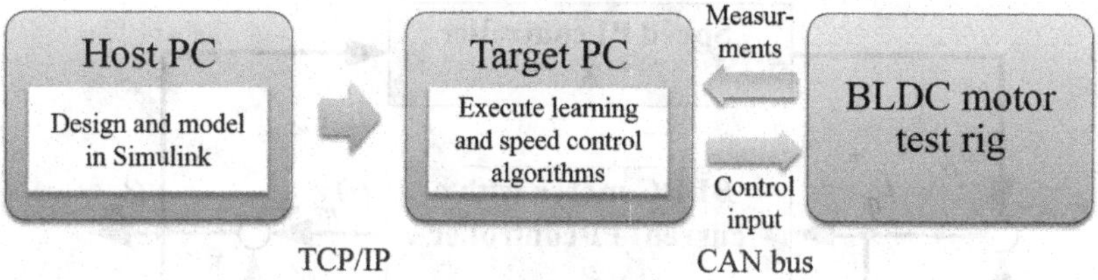

Figure 5.17 Components of the xPC target environment

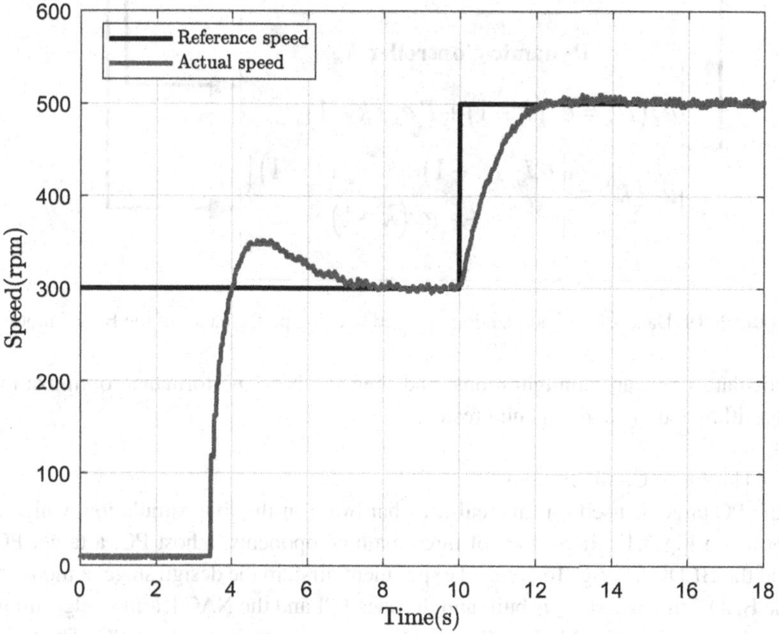

Figure 5.18 Tracking performance of the BLDC motor system with a speed PI controller

current controllers are designed and embedded in a motor electronic control unit (ECU). The experiment is then performed in two steps: 1) to contrive performance degradation, we design a speed PI controller K to achieve closed-loop stability but with a poor speed tracking performance, 2) to enhance the performance, we apply the proposed NAC learning algorithm. To be specific, in what follows, we will first

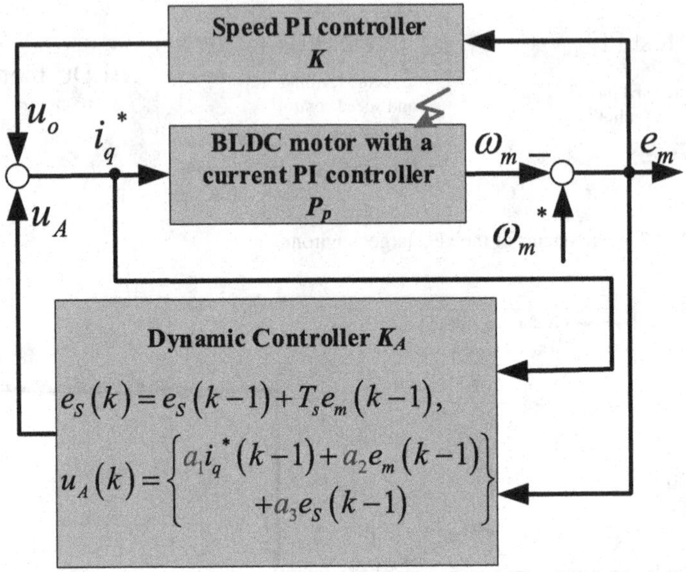

Figure 5.19 Data-driven optimization of speed tracking performance of the BLDC motor

articulate hardware configurations, and then the NAC performance optimization algorithm and the experimental results.

(A). Hardware Configurations
An xPC target is used for this real-time hardware-in-the-loop simulation, which is shown in Fig. 5.17. It consists of three main components, a host PC, a target PC, and the BLDC test rig. To conduct experiment, first, in the design stage, a model of the BLDC motor test rig is built and the speed PI and the NAC learning algorithms are developed in the MATLAB/Simulink environment in the host PC. Then, the algorithms are compiled and sent to the target PC through a TCP/IP protocol. And the target PC is in a ready-to-work mode. Once the BLDC motor test rig is powered on and the target PC receives a trigger signal from the host PC, it starts receiving measurments from the BLDC motor test rig and sending back control signals via a CAN bus in real time.

(B). The NAC Performance Optimization Algorithm and the Experimental Results
In the BLDC motor test rig, the accessible input variable is the reference current in the q-axis i_q^*, and the measurable output variable is the mechanical motor speed

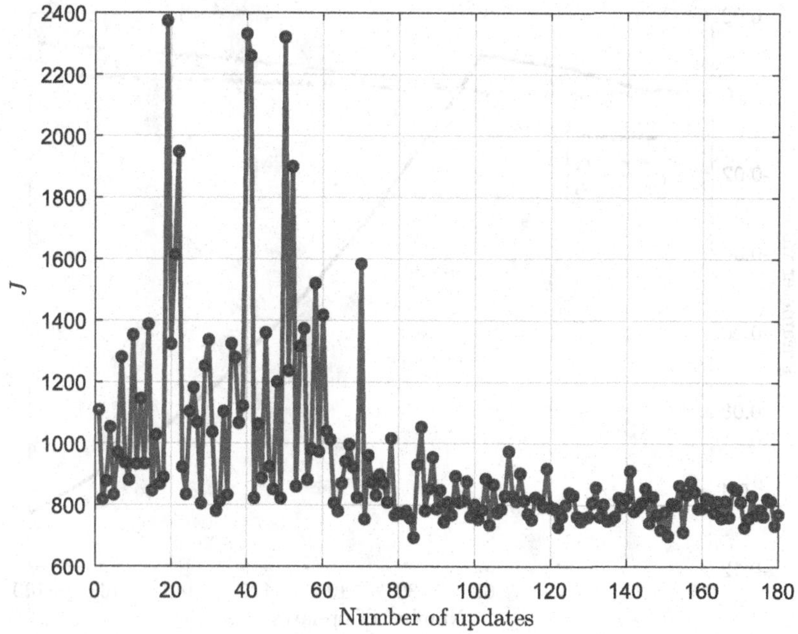

Figure 5.20 Cost J during online learning

ω_m. Initially, a speed PI controller K is designed to track the reference speed ω_m^*. To check the closed-loop tracking performance, an experiment is performed by setting the reference speed $\omega_m^* = 300\varepsilon(t) + 200\varepsilon(t-10)$ rpm, where $\varepsilon(t)$ is the unit step function, and the sampling time $T_s = 10$ ms. The speed response is shown in Fig. 5.18. It is clear that the system has a long settling time and a large overshoot. Additionally, it shows clear nonlinearity. In order to improve the tracking performance, the proposed NAC method in Subsection 5.4.2 is applied. The following value function is chosen

$$\text{Minimize} \quad V(k) = \lim_{N\to\infty} \mathbb{E}\left(\sum_{i=k}^{N} \gamma^{i-k}\left(i_q^{*T}(i)Ri_q^*(i) + e_S^T(i)Qe_S(i)\right)\right),$$

$$\text{subject to} \quad J(j) = \sum_{k=Hj}^{H(j+1)-1} \gamma^{k-Hj}\left(i_q^{*T}(k)Ri_q{}^*(k) + e_S^T(k)Qe_S(k)\right) \le J_{th},$$

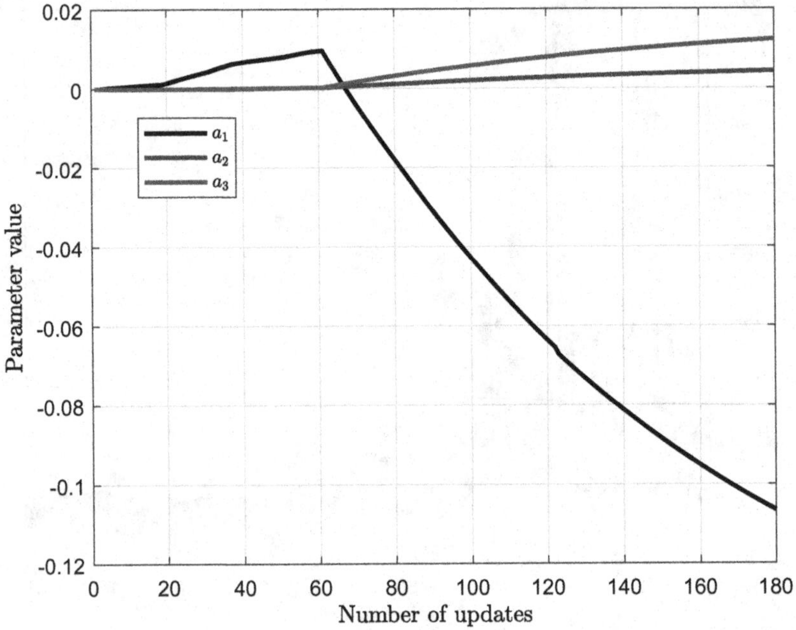

Figure 5.21 Parameters a_1, a_2 and a_3 during online learning

where e_S denotes the sum of speed tracking errors e_m that is computed by $e_S(k+1) = e_S(k) + T_s e_m(k)$. It is also the sole state of the speed PI controller K. The weighting factors are set to $R = 1$ and $Q = 0.001$, and the discount factor γ is set to 0.99. The time horizon H of the evaluation function J is set to 500 (this equals 5 s) and its threshold J_{th} is 3500 for stability detection.

The controller K_A is selected as a first-order controller that is parameterized by

$$u_A(k) = a_1 i_q^*(k-1) + a_2 e_m(k-1) + a_3 e_S(k),$$

where a_1, a_2, and a_3 are parameters to be optimized. The length of the past I/O data, i_q^* and e_m, used for the parameterization of the value function (5.42) is set to 8. The whole control diagram is shown in Fig. 5.19. For learning, we first maintain the speed at 500 rpm and then trigger the learning procedure. Fig. 5.20 and Fig. 5.21 show the change of cost and parameters of K_A during online learning. Seen from the gradual decreasing cost in Fig. 5.20, the updating directions of the parameters

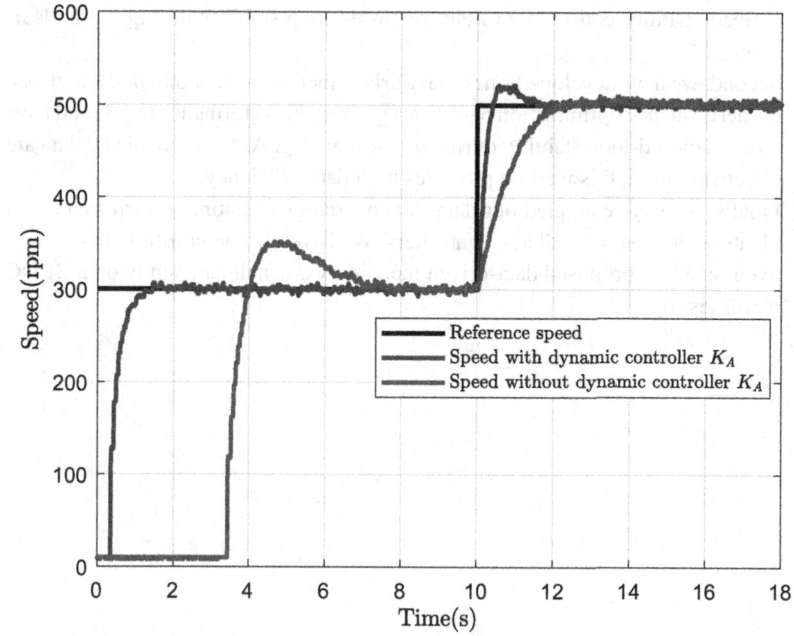

Figure 5.22 Comparison of performance of systems with and without K_A

of K_A are correct. Finally, after 180 updates, the learning procedure is stopped. The learned K_A is then placed back into the system. Fig. 5.22 shows the comparison of system performance with and without the learned K_A. It can be seen from Fig. 5.22 that, with the learned K_A the system shows considerably less overshoot and settling time. This demonstrates the effectiveness of the proposed performance optimization method.

5.5 Concluding Remarks

In this chapter, we have made three key contributions to solving performance optimization problems in a stochastic setting.

- First, we have derived conditions for a successful application of the on-policy NAC method to the optimization of robustness against plant perturbations, and

derived stability constraints to ensure closed-loop stability during online learning.
- Second, we have developed a new data-driven method using both model and data for performance optimization based on a prescribed performance index, and have ensured closed-loop stability during online learning. As both model and data are effectively used, this method promises high data-efficiency.
- Finally, we have extended our data-driven strategy to optimize performance of plants with general feedback controllers. We have also demonstrated the effectiveness of the proposed data-driven method by a benchmark study on a BLDC motor test rig.

Conclusion and Future Work 6

The greater goal of the thesis has been the establishment of a RL framework for performance optimization of industrial systems. Several significant contributions have been made to bring us one step closer to the ultimate aim. In this chapter, we first conclude the main insights and contributions of the work, and then point out new directions, in which further research can be conducted.

6.1 Conclusion

This thesis has been dedicated to the development of RL aided methods for performance optimization of feedback control systems. It integrated theoretical fundamentals of RL and feedback control, and delivered new and practical results that could benefit the modern industry. The thesis has been concluded as follows.

In Chapter 1, we have given a brief introduction of the work. It started with the motivation of the chosen topic, i.e. *"Reinforcement learning aided performance optimization of feedback control systems"* and the research scope of the work in the context of implementation of performance optimization in industrial systems. It included the objective of the work and an outline of the remainder of the thesis. In Chapter 2, we have described the design of feedback controllers for a nominal LTI plant and the analysis of stability of closed-loop systems with uncertainty. This provides preliminaries about feedback control.

In Chapter 3, we have presented the most extensive review of DP and RL methods for feedback control as well as a novel RL approach for it. We started out by giving an overview of the state-of-the-art RL methods by classifying them into three categories, i.e. actor-only, critic-only and actor-critic. We have derived two DP-based methods, policy iteration and value iteration based on the Bellman equation and Bellman optimality equation, and applied them to solve the LQR and LQG

C. Hua, *Reinforcement Learning Aided Performance Optimization of Feedback Control Systems*, https://doi.org/10.1007/978-3-658-33034-7_6

problems using DP. We have further derived two critic-only RL methods, Q-learning and SARSA, to solve the LQR problem. To deal with the LQG problem, we have proposed a new NAC method for the optimization of a dynamic output feedback controller with a prescribed structure. This method has been, to the best knowledge of the author, the first try of efficiently solving an optimization problem in a partially observable and stochastic environment.

In Chapter 4, we have presented IOR and performance index based approaches, for performance optimization of plants with observer-based state feedback controllers, where a residual vector is available. The proposed IOR method, unlike many existing robustness optimization approaches, recovers the I/O responses of the perturbed system to those of an ideal system consisting of the nominal plant and an LQR. This enables system robustness to be optimized in a data-driven manner. In addition, as the latter system has good performance and robustness guarantees, this IOR strategy promises good optimization results. On the other hand, the proposed performance index based approach recovers the performance of the perturbed system according to the performance index, based on which the existing controller is built. We have also extended the performance index based approach to solve a tracking performance degradation problem. Finally, we have implemented all performance optimization methods by Q-learning in a deterministic environment, which shows successful applications of RL methods to performance optimization of feedback control systems.

Chapter 5 was dedicated to the data-driven implementation of the IOR and performance index based approaches proposed in Chapter 4 in a stochastic environment. We have, first, derived conditions for the use of the on-policy NAC method proposed in Chapter 3 for the data-driven implementation of the IOR method. We have also derived conditions to ensure the stability of the closed-loop system during online learning. Subsequently, we have developed a data-efficient learning approach using both model and data for the implementation of the performance index based method. Finally, we have extended the performance index based method to optimize the performance of plants with general feedback controllers, such as PID controllers, and demonstrated its effectiveness on a BLDC motor test rig.

In summary, in this thesis, we have mainly proposed

- a NAC approach to deal with the LQG problem (Chapter 3).
- an IOR approach and a performance index based approach to improve the robustness and performance of perturbed systems with observer-based state feedback controllers (Chapter 4).
- Q-learning methods for data-driven implementation of the above approaches in a deterministic environment (Chapter 4).

- NAC methods for data-driven implementation of the above approaches in a stochastic environment (Chapter 5).
- a NAC method for data-driven implementation of the performance index based approach to improve performance of perturbed systems with general output feedback controllers, which works well on a BLDC motor test rig (Chapter 5).

6.2 Future Work

While this thesis made great contributions to performance optimization of feedback control systems using RL methods, there are some aspects we have not included but worth being explored.

Performance Monitoring. In Chapter 1, we have pointed out that for successful performance optimization of feedback control systems, two procedures can be performed: performance monitoring and controller reconfiguration. So far, we have only covered the latter procedure. Therefore, performance monitoring methods [33, 36] should be developed. From the viewpoint of RL, the TD error, which was extensively used in the development of performance optimization methods in this thesis, can be used to quantify the level of the system performance degradation after perturbations occur. The main issues are:

- the determination of an optimal evaluation function based on the TD error;
- the determination of a threshold according to a given false alarm rate.

It is promising to further build a general RL framework for performance monitoring/optimization of modern industrial systems by merging the above methods with the optimization methods proposed in this thesis.

Sample-efficient NAC Methods. In Chapter 3, to deal with the LQG problem, we have presented a new on-policy NAC learning method that enabled us to online learn an optimal output feedback controller. However, the NAC learning method itself is not sample-efficient, as it uses only each data point once. In comparison, off-policy RL methods are very powerful in data reuse, which have high data efficiency but often with a sacrifice of algorithm stability [27, 65]. In light of this, most recent studies have been carried out in the computer science community to bring together on- and off-policy methods with the hope of gaining advantages from both modes of learning, while avoiding their limitations [26, 75]. Based on these studies, it is of great interests to develop more sample-efficient NAC methods for solving the

LQG problem, and to further apply them to deal with performance optimization problems.

Learning of Optimal Controllers for Performance Optimization. In Chapter 4, we have found out controllers for performance optimization of deterministic systems by Q-learning. We have pointed out that these controllers consisting of deadbeat observers and optimal state feedback gains, are suboptimal, whereas an optimal controller should be a $\mathcal{H}_2$ controller consisting of a $\mathcal{H}_2$ observer and an optimal state feedback gain. Likewise, in Chapter 5, we have found out controllers with prescribed structures for performance optimization of stochastic systems using the proposed NAC methods. They are also suboptimal, because an optimal controller should be an LQG controller consisting of a Kalman observer and an optimal state feedback gain. Therefore, it would be valuable to develop methods that allow optimal controllers to be learned for performance optimization of both deterministic and stochastic systems.

Performance Optimization of Nonlinear Systems. All RL aided performance optimization methods developed in this work are only suitable for LTI systems. This can restrict the use of our methods in industrial systems. To deal with nonlinear systems operating around an equilibrium point, linearization of the systems around that point allows all developed RL methods to be used for performance optimization. However, if the goal is to track a given trajectory, then the developed methods cannot be applied. It is known that differential dynamic programming is a model-based local dynamic programming algorithm that can be used in nonlinear systems for trajectory planning [35, 70]. Therefore, it would be interesting to establish the connection between the proposed RL aided methods and differential dynamic programming, so that the optimization of tracking performance of nonlinear systems can be achieved. Besides, it would also be of practical interests to investigate general RL aided methods for performance optimization of nonlinear systems.

Bibliography

1. A. Al-Tamimi, F. L. Lewis, and M. Abu-Khalaf, "Model-free Q-learning designs for linear discrete-time zero-sum games with application to H_∞ control," *Automatica*, vol. 43, no. 3, pp. 473–481, 2007.
2. S.-I. Amari, "Natural gradient works efficiently in learning," *Neural computation*, vol. 10, no. 2, pp. 251–276, 1998.
3. B. D. Anderson, "From Youla-Kučera to identification, adaptive and nonlinear control," *Automatica*, vol. 34, no. 12, pp. 1485–1506, 1998.
4. M. Athans, "A tutorial on the LQG/LTR method," in *Proceedings of the 1986 American Control Conference*. IEEE, 1986, pp. 1289–1296.
5. R. Bellman, *Dynamic programming*. Princeton, NJ: Princeton Univ. Press, 1957.
6. J. Bendtsen, K. Trangbaek, and J. Stoustrup, "Plug-and-play control—Modifying control systems online," *IEEE Transactions on Control Systems Technology*, vol. 21, no. 1, pp. 79–93, 2011.
7. D. P. Bertsekas and A. Scientific, "Reinforcement learning and optimal control," *Athena Scientific, Belmont, Massachusetts*, 2019.
8. D. P. Bertsekas and J. N. Tsitsiklis, *Neuro-dynamic programming*. Athena Scientific Belmont, MA, 1996, vol. 5.
9. M. Blanke, M. Kinnaert, J. Lunze, M. Staroswiecki, and J. Schröder, *Diagnosis and fault-tolerant control*, 2nd ed. Springer, 2006.
10. S. J. Bradtke, B. E. Ydstie, and A. G. Barto, "Adaptive linear quadratic control using policy iteration," in *Proceedings of the 1994 American Control Conference*, vol. 3. IEEE, 1994, pp. 3475–3479.
11. M.-J. Chen and C. A. Desoer, "Necessary and sufficient condition for robust stability of linear distributed feedback systems," *International Journal of Control*, vol. 35, no. 2, pp. 255–267, 1982.
12. C. Desoer and M. Vidyasagar, "Feedback systems: Input-output properties," *Society for Industrial Mathematics*, 2009.
13. S. X. Ding, *Model-based fault diagnosis techniques: design schemes, algorithms, and tools*, 2nd ed. London: Springer-Verlag, 2013.
14. S. X. Ding, *Data-driven design of fault diagnosis and fault-tolerant control systems*. London: Springer-Verlag, 2014.
15. S. X. Ding, *Advanced methods of fault diagnosis and fault-tolerant control*. London: Springer-Verlag, 2020.

C. Hua, *Reinforcement Learning Aided Performance Optimization of Feedback Control Systems*, https://doi.org/10.1007/978-3-658-33034-7

16. J. Doyle and G. Stein, "Multivariable feedback design: Concepts for a classical/modern synthesis," *IEEE Transactions on Automatic Control*, vol. 26, no. 1, pp. 4–16, 1981.
17. J. Doyle and G. Stein, "Robustness with observers," *IEEE Transactions on Automatic Control*, vol. 24, no. 4, pp. 607–611, 1979.
18. J. C. Doyle, "Guaranteed margins for LQG regulators," *IEEE Transactions on Automatic Control*, vol. 23, no. 4, pp. 756–757, 1978.
19. J. C. Doyle, B. A. Francis, and A. R. Tannenbaum, *Feedback control theory*. Courier Corporation, 2013.
20. C. Finn, S. Levine, and P. Abbeel, "Guided cost learning: Deep inverse optimal control via policy optimization," in *International Conference on Machine Learning*, 2016, pp. 49–58.
21. B. A. Francis, *A course in H_∞ control theory*. Berlin; New York: Springer-Verlag, 1987.
22. T. T. Georgiou and M. C. Smith, "Optimal robustness in the gap metric," *IEEE Transactions on Automatic Control*, vol. 35, no. 6, pp. 673–686, 1990.
23. G. C. Goodwin and K. S. Sin, *Adaptive filtering prediction and control*. Courier Corporation, 2014.
24. M. Green and D. J. Limebeer, *Linear robust control*. Courier Corporation, 2012.
25. I. Grondman, L. Busoniu, G. A. Lopes, and R. Babuska, "A survey of actor-critic reinforcement learning: Standard and natural policy gradients," *IEEE Transactions on Systems, Man, and Cybernetics, Part C (Applications and Reviews)*, vol. 42, no. 6, pp. 1291–1307, 2012.
26. S. Gu, T. Lillicrap, Z. Ghahramani, R. E. Turner, and S. Levine, "Q-prop: Sample-efficient policy gradient with an off-policy critic," *arXiv preprint:1611.02247*, 2016.
27. S. S. Gu, T. Lillicrap, R. E. Turner, Z. Ghahramani, B. Schölkopf, and S. Levine, "Interpolated policy gradient: Merging on-policy and off-policy gradient estimation for deep reinforcement learning," in *Advances in Neural Information Processing Systems*, 2017, pp. 3846–3855.
28. V. Gullapalli, J. A. Franklin, and H. Benbrahim, "Acquiring robot skills via reinforcement learning," *IEEE Control Systems Magazine*, vol. 14, no. 1, pp. 13–24, 1994.
29. F. Hansen, "Fractional representation approach to closed loop identification and experiment design," Ph.D. dissertation, Department of Electrical Engineering, Standford University, 1989.
30. F. Hansen, G. Franklin, and R. Kosut, "Closed-loop identification via the fractional representation: Experiment design," in *Proceedings of the 1989 American Control Conference*. IEEE, 1989, pp. 1422–1427.
31. P. He and S. Jagannathan, "Reinforcement learning neural-network-based controller for nonlinear discrete-time systems with input constraints," *IEEE Transactions on Systems, Man, and Cybernetics, Part B (Cybernetics)*, vol. 37, no. 2, pp. 425–436, 2007.
32. C. Hua, S. X. Ding, and Y. A. Shardt, "A new method for fault tolerant control through Q-learning," *IFAC-PapersOnLine*, vol. 51, no. 24, pp. 38–45, 2018.
33. B. Huang and S. L. Shah, *Performance assessment of control loops: theory and applications*. Springer Science & Business Media, 1999.
34. A. J. Ijspeert, J. Nakanishi, and S. Schaal, "Learning attractor landscapes for learning motor primitives," in *Advances in Neural Information Processing Systems*, 2003, pp. 1547–1554.
35. D. H. Jacobson and D. Q. Mayne, "Differential dynamic programming," 1970.

36. M. Jelali, "Control system performance monitoring: Assessment, diagnosis and improvement of control loop performance in industrial automation," Ph.D. dissertation, 2010.
37. S. M. Kakade, "A natural policy gradient," in *Advances in Neural Information Processing Systems*, 2002, pp. 1531–1538.
38. J. Kober, J. A. Bagnell, and J. Peters, "Reinforcement learning in robotics: A survey," *The International Journal of Robotics Research*, vol. 32, no. 11, pp. 1238–1274, 2013.
39. J. Kober and J. R. Peters, "Policy search for motor primitives in robotics," in *Advances in Neural Information Processing Systems*, 2009, pp. 849–856.
40. V. Kučera, *Discrete linear control: the polynomial equation approach*. J. Wiley Chichester, 1979.
41. F. L. Lewis and D. Vrabie, "Reinforcement learning and adaptive dynamic programming for feedback control," *IEEE circuits and Systems Magazine*, vol. 9, no. 3, pp. 32–50, 2009.
42. F. L. Lewis, D. Vrabie, and K. G. Vamvoudakis, "Reinforcement learning and feedback control: Using natural decision methods to design optimal adaptive controllers," *IEEE Control Systems Magazine*, vol. 32, no. 6, pp. 76–105, 2012.
43. L. Li, H. Luo, S. X. Ding, Y. Yang, and K. Peng, "Performance-based fault detection and fault-tolerant control for automatic control systems," *Automatica*, vol. 99, pp. 308–316, 2019.
44. M. Liu and P. Shi, "Sensor fault estimation and tolerant control for Itô stochastic systems with a descriptor sliding mode approach," *Automatica*, vol. 49, no. 5, pp. 1242–1250, 2013.
45. L. Ljung, "System identification," *Wiley Encyclopedia of Electrical and Electronics Engineering*, pp. 1–19, 1999.
46. H. Luo, *Plug-and-play Monitoring and Performance Optimization for Industrial Automation Processes*. Springer, 2017.
47. H. Luo, X. Yang, M. Krueger, S. X. Ding, and K. Peng, "A plug-and-play monitoring and control architecture for disturbance compensation in rolling mills," *IEEE/ASME Transactions on Mechatronics*, vol. 23, no. 1, pp. 200–210, 2016.
48. V. Mnih, A. P. Badia, M. Mirza, A. Graves, T. Lillicrap, T. Harley, D. Silver, and K. Kavukcuoglu, "Asynchronous methods for deep reinforcement learning," in *International Conference on Machine Learning*, 2016, pp. 1928–1937.
49. V. Mnih, K. Kavukcuoglu, D. Silver, A. A. Rusu, J. Veness, M. G. Bellemare, A. Graves, M. Riedmiller, A. K. Fidjeland, G. Ostrovski *et al.*, "Human-level control through deep reinforcement learning," *Nature*, vol. 518, no. 7540, pp. 529–532, 2015.
50. J. Moore and T. Tay, "Loop recovery via H_∞/H_2 sensitivity recovery," *International Journal of Control*, vol. 49, no. 4, pp. 1249–1271, 1989.
51. H. Niemann, "Dual youla parameterisation," *IEE Proceedings-Control Theory and Applications*, vol. 150, no. 5, pp. 493–497, 2003.
52. H. Nyquist, "Regeneration theory," *Bell System Technical Journal*, vol. 11, no. 1, pp. 126–147, 1932.
53. J. Peters, "Machine learning of motor skills for robotics," Ph.D. dissertation, University of Southern California, 2007.
54. J. Peters and S. Schaal, "Natural actor-critic," *Neurocomputing*, vol. 71, no. 7–9, pp. 1180–1190, 2008.

55. J. Peters, S. Vijayakumar, and S. Schaal, "Natural actor-critic," in *European Conference on Machine Learning*. Springer, 2005, pp. 280–291.
56. S. J. Qin, "An overview of subspace identification," *Computers & chemical engineering*, vol. 30, no. 10–12, pp. 1502–1513, 2006.
57. M. Safonov and M. Athans, "Gain and phase margin for multiloop LQG regulators," *IEEE Transactions on Automatic Control*, vol. 22, no. 2, pp. 173–179, 1977.
58. J. Schulman, "Optimizing expectations: From deep reinforcement learning to stochastic computation graphs," Ph.D. dissertation, UC Berkeley, 2016.
59. J. Schulman, S. Levine, P. Abbeel, M. Jordan, and P. Moritz, "Trust region policy optimization," in *International Conference on Machine Learning*, 2015, pp. 1889–1897.
60. J. Schulman, F. Wolski, P. Dhariwal, A. Radford, and O. Klimov, "Proximal policy optimization algorithms," *arXiv preprint:1707.06347*, 2017.
61. R. Shadmehr and S. Mussa-Ivaldi, *Biological learning and control: how the brain builds representations, predicts events, and makes decisions*. MIT Press, 2012.
62. D. Silver, A. Huang, C. J. Maddison, A. Guez, L. Sifre, G. Van Den Driessche, J. Schrittwieser, I. Antonoglou, V. Panneershelvam, M. Lanctot *et al.*, "Mastering the game of go with deep neural networks and tree search," *Nature*, vol. 529, no. 7587, pp. 484–489, 2016.
63. D. Silver, G. Lever, N. Heess, T. Degris, D. Wierstra, and M. Riedmiller, "Deterministic policy gradient algorithms," in *International Conference on Machine Learning*, 2014.
64. D. Silver, J. Schrittwieser, K. Simonyan, I. Antonoglou, A. Huang, A. Guez, T. Hubert, L. Baker, M. Lai, A. Bolton *et al.*, "Mastering the game of go without human knowledge," *Nature*, vol. 550, no. 7676, pp. 354–359, 2017.
65. R. S. Sutton and A. G. Barto, *Reinforcement learning: An introduction*, 2nd ed. MIT press, 2018.
66. R. S. Sutton, D. A. McAllester, S. P. Singh, and Y. Mansour, "Policy gradient methods for reinforcement learning with function approximation," in *Advances in Neural Information Processing Systems*, 2000, pp. 1057–1063.
67. I. Szita and A. Lörincz, "Learning tetris using the noisy cross-entropy method," *Neural Computation*, vol. 18, no. 12, pp. 2936–2941, 2006.
68. T.-T. Tay, I. Mareels, and J. B. Moore, *High performance control*. Springer Science & Business Media, 1998.
69. T. Tay, J. Moore, and R. Horowitz, "Indirect adaptive techniques for fixed controller performance enhancement," *International Journal of Control*, vol. 50, no. 5, pp. 1941–1959, 1989.
70. E. Theodorou, Y. Tassa, and E. Todorov, "Stochastic differential dynamic programming," in *Proceedings of the 2010 American Control Conference*. IEEE, 2010, pp. 1125–1132.
71. E. Todorov, "Stochastic optimal control and estimation methods adapted to the noise characteristics of the sensorimotor system," *Neural Computation*, vol. 17, no. 5, pp. 1084–1108, 2005.
72. M. Vidyasagar, "The graph metric for unstable plants and robustness estimates for feedback stability," *IEEE Transactions on Automatic Control*, vol. 29, no. 5, pp. 403–418, 1984.
73. M. Vidyasagar, "Control system synthesis: a factorization approach, part II," *Synthesis Lectures on Control and Mechatronics*, vol. 2, no. 1, pp. 1–227, 2011.

74. K. Wampler and Z. Popović, "Optimal gait and form for animal locomotion," in *ACM Transactions on Graphics (TOG)*, vol. 28, no. 3. ACM, 2009, pp. 60:1–60:8.
75. Z. Wang, V. Bapst, N. Heess, V. Mnih, R. Munos, K. Kavukcuoglu, and N. de Freitas, "Sample efficient actor-critic with experience replay," *arXiv preprint:1611.01224*, 2016.
76. C. J. Watkins, "Learning from delayed rewards," Ph.D. dissertation, King's college, 1989.
77. D. Wierstra, T. Schaul, J. Peters, and J. Schmidhuber, "Natural evolution strategies," in *2008 IEEE Congress on Evolutionary Computation*. IEEE, 2008, pp. 3381–3387.
78. R. J. Williams, "Simple statistical gradient-following algorithms for connectionist reinforcement learning," *Machine Learning*, vol. 8, no. 3–4, pp. 229–256, 1992.
79. W.-Y. Yan and J. B. Moore, "Stable linear fractional transformations with applications to stabilization and multistage H_∞ control design," *International Journal of Robust and Nonlinear Control*, vol. 6, no. 2, pp. 101–122, 1996.
80. D. Youla, J. d. Bongiorno, and H. Jabr, "Modern Wiener–Hopf design of optimal controllers Part I: The single-input-output case," *IEEE Transactions on Automatic Control*, vol. 21, no. 1, pp. 3–13, 1976.
81. D. Youla, H. Jabr, and J. Bongiorno, "Modern Wiener-Hopf design of optimal controllers–Part II: The multivariable case," *IEEE Transactions on Automatic Control*, vol. 21, no. 3, pp. 319–338, 1976.
82. Y. Zhang and J. Jiang, "Fault tolerant control system design with explicit consideration of performance degradation," *IEEE Transactions on Aerospace and Electronic Systems*, vol. 39, no. 3, pp. 838–848, 2003.
83. Y. Zhang and J. Jiang, "Bibliographical review on reconfigurable fault-tolerant control systems," *Annual Reviews in Control*, vol. 32, no. 2, pp. 229–252, 2008.
84. D. Zhao, H. Wang, K. Shao, and Y. Zhu, "Deep reinforcement learning with experience replay based on SARSA," in *2016 IEEE Symposium Series on Computational Intelligence (SSCI)*. IEEE, 2016, pp. 1–6.
85. K. Zhou, J. C. Doyle, K. Glover *et al.*, *Robust and optimal control*. Prentice hall New Jersey, 1996.

Printed in the United States
By Bookmasters